797,885 Books

are available to read at

Forgotten Books

www.ForgottenBooks.com

Forgotten Books' App
Available for mobile, tablet & eReader

ISBN 978-1-331-94084-5
PIBN 10257433

This book is a reproduction of an important historical work. Forgotten Books uses state-of-the-art technology to digitally reconstruct the work, preserving the original format whilst repairing imperfections present in the aged copy. In rare cases, an imperfection in the original, such as a blemish or missing page, may be replicated in our edition. We do, however, repair the vast majority of imperfections successfully; any imperfections that remain are intentionally left to preserve the state of such historical works.

Forgotten Books is a registered trademark of FB &c Ltd.
Copyright © 2015 FB &c Ltd.
FB &c Ltd, Dalton House, 60 Windsor Avenue, London, SW19 2RR.
Company number 08720141. Registered in England and Wales.

For support please visit www.forgottenbooks.com

1 MONTH OF FREE READING

at

www.ForgottenBooks.com

By purchasing this book you are eligible for one month membership to ForgottenBooks.com, giving you unlimited access to our entire collection of over 700,000 titles via our web site and mobile apps.

To claim your free month visit:
www.forgottenbooks.com/free257433

* Offer is valid for 45 days from date of purchase. Terms and conditions apply.

English
Français
Deutsche
Italiano
Español
Português

www.forgottenbooks.com

Mythology Photography **Fiction**
Fishing Christianity **Art** Cooking
Essays Buddhism Freemasonry
Medicine **Biology** Music **Ancient Egypt** Evolution Carpentry Physics
Dance Geology **Mathematics** Fitness
Shakespeare **Folklore** Yoga Marketing
Confidence Immortality Biographies
Poetry **Psychology** Witchcraft
Electronics Chemistry History **Law**
Accounting **Philosophy** Anthropology
Alchemy Drama Quantum Mechanics
Atheism Sexual Health **Ancient History**
Entrepreneurship Languages Sport
Paleontology Needlework Islam
Metaphysics Investment Archaeology
Parenting Statistics Criminology
Motivational

LONDON AGENTS
SIMPKIN, MARSHALL, HAMILTON, KENT & Co. Ltd.

THE AUTONOMIC NERVOUS SYSTEM

BY

J. N. LANGLEY,

Sc.D., Hon. LL.D., Hon. M.D., F.R.S.

Professor of Physiology in the University of Cambridge

PART I.

CAMBRIDGE
W. HEFFER & SONS LTD.
1921

Contents

PAGE

1. THE DIVISIONS OF THE AUTONOMIC SYSTEM. NOMENCLATURE - - - - -
    Notes.—NOMENCLATURE. VOLUNTARY PRODUCTION OF AUTONOMIC EFFECTS, p. 12.

2. GENERAL PLAN OF ORIGIN AND OF PERIPHERAL DISTRIBUTION - - - - 15

3. THE NERVE FIBRES OF THE AUTONOMIC SYSTEM - - - - - - 22

4. THE SPECIFIC ACTION OF DRUGS ON THE SYMPATHETIC AND PARASYMPATHETIC SYSTEMS 28

    a. ADRENALINE AND SYMPATHETIC EFFECTS 29
    Notes.—FALL OF BLOOD PRESSURE AND DILATATION OF BLOOD VESSELS; ACTION ON CEREBRAL VESSELS, VEINS, CAPILLARIES, p. 31. ACTION ON MUSCLE, p. 33.

    b. PILOCARPINE AND PARASYMPATHETIC EFFECTS - - - - - 35
    Notes.—ACTION ON UTERUS AND BLOOD VESSELS, p. 36.

    c. REVERSAL OF EFFECT OF SYMPATHETIC AND PARASYMPATHETIC DRUGS 37

    d. THEORIES ON THE RELATION OF DRUGS TO NERVE SYSTEMS - - - - 39
    Note.—LOCALISATION OF ACTION OF DRUGS ON STRIATED MUSCLE, p. 47.

    e. CLASSIFICATION OF SYMPATHETIC AND PARASYMPATHETIC NERVES ACCORDING TO PHARMACOLOGICAL RESULTS 50
    Notes.—HISTORICAL. PARASYMPATHOTONIA SYMPATHOTONIA. PERITERMINAL NETWORK, p. 53.

5. THE TISSUES INNERVATED - - - 57

    a. PIGMENT CELLS - - - - - 59

    b. SYMPATHETIC NERVES AND THE INNERVATION OF CAPILLARIES - - - 64
    Notes.—CAPILLARY CONTRACTILITY, p. 67.

    c. SYMPATHETIC NERVES AND THE INNERVATION OF STRIATED MUSCLE 69

Some Abbreviations in Reference to Papers

A.	−Archiv, -es; Archivio	Nervenhk.	=Nervenheilkunde.
		néerl.	=néerlandaise.
Anz.	=Anzeiger.	Neurob.	=Neurobiologica.
Berl.	−Berlin.	Proc.	−Proceedings.
Bull.	=Bulletin.	Quar.	−Quarterly.
Cntrlb.	=Centralblatt.	Rep.	−Reports.
Comp.	=Comparative.	Sitzb.	=Situngsberichte.
C.R.	=Comptes rendus.	Sch.	=Schola.
deut.	=deutsche, -es.	Ther.	=Therapeutics, Therapie.
Ergeh.	=Ergebnisse.		
ges.	=gesammte.	Trans.	=Transactions.
J.	=Journal.	Verh.	=Verhandlungen.
J.H.	=Johns Hopkins.	wiss.	=wissenschaftliche.
Mik.	=Mikroscopische.	Wochens.	=Wochenschrift.
Monats.	−Monatschrift.	Zntrlb.	−Zentralblatt.
Münch.	=Münchener.	Zts.	=Zeitschrift.

1. The Divisions of the Autonomic System. Nomenclature.

The autonomic nervous system consists of the nerve cells and nerve fibres, by means of which efferent impulses pass to tissues other than multi-nuclear striated muscle.

The progress of knowledge of this system has been continuous for at least the last 250 years, though at times it has progressed but slowly. As knowledge increased, new points of view presented themselves, and the terms used to express the general conceptions naturally varied. I give a short historical account of these terms, omitting, as far as practicable, reference to the actual state of knowledge at the successive periods.

VOLUNTARY AND INVOLUNTARY. In the early part of the 18th century the movements of the various parts of the body were commonly divided into three classes, viz. (1) voluntary movements, (2) involuntary movements which could also be produced by the will, such as the movements of the respiratory muscles in sleep and instinctive movements, (3) involuntary movements over which the will had little or no control, such as the movements of the heart and intestines; this form of involuntary movement was called vital or natural movement.

In later times, in such observations as were mainly confined to the vital movements, the nerves supposed to be concerned with them were not infrequently spoken of as involuntary nerves, but in a general classification, a division into voluntary and involuntary nerves was rarely adopted. The use of "involuntary" in more or less of the sense in which we have used the term autonomic occurs however throughout the 19th century, and was so used by Gaskell in his last work in 1914.

THE SYMPATHETIC NERVES. A different terminology began with Winslow (1732). The nerve called the intercostal nerve was known to be the chief nerve to send fibres to the intestines, and it was known that it also sent fibres to the heart. In consequence of its numerous connexions it was supposed to be the means by which one part of the body affected another part, or, as it was then stated, by which the sympathies of the body were brought about. Winslow, in view of this, called the nerve the great sympathetic. Two other nerves Winslow considered had a similar function, viz., the portio dura of the seventh nerve, which he called the small sympathetic, and the vagus, which he called the medium sympathetic.

The terms small sympathetic and medium sympathetic were rarely used. In France "great sympathetic" remained as the designation of the intercostal nerve; in other countries, and sometimes in France, "great" was dropped, and the intercostal nerve became the sympathetic nerve, and with its branches, the sympathetic nervous system. The meaning of sympathetic nervous system has from time to time been extended to include more or less of the rest of the autonomic system.

GANGLIONIC NERVES AND GANGLIONIC NERVOUS SYSTEM. A third terminology arose from the investigation of the ganglia on the course of the nerves.

Johnstone (1764) put forward the theory that the ganglia served to convert voluntary into involuntary movements. Thus the nerves governing involuntary movements could naturally be spoken of as ganglionic nerves. The ganglionic nerves were not confined to the nerves governing vital movements, but included the nerves governing the involuntary movements of muscles which could also be put in action by the will, and thus the ganglia on the course of the cranial and spinal nerves were included in the ganglionic nervous system. In the course of time most of the cells of these ganglia were recognised as belonging to afferent nerves, and were excluded from the ganglionic system, but the term was rarely, if ever, used to exclude them all.

NERVOUS SYSTEM OF ANIMAL AND OF ORGANIC LIFE. Additional terms were adopted by Bichat (1800-1). On the lines of suggestions made by some earlier writers, he divided the life of the organism into the animal life (vie de relation) and the vegetative or organic life (vie de nutrition). There were thus vegetative and ammal nerves. The ganglia of the great sympathetic and some of those of the cranial nerves, he said—with various reservations and obscurities—were the nervous organs of organic life. The theory was much discussed, but was either rejected or modified by most of those who submitted it to the test of experiment. It increased for a time the use of the terms ganglionic nerves and ganglionic nervous system, and not infrequently led to the use of the terms organic nerves and organic nervous system. The theory and the term organic nearly died out about the middle of the 19th century. "Vegetative" has recently been revived as a substitute for "autonomic."

CEREBRO-SPINAL AND SYMPATHETIC NERVES. In the first half of the 19th century, concurrently with the use of the terms mentioned above, it was not uncommon to divide the nerves into cerebro-spinal nerves and their ganglia on the one hand, and sympathetic nerves and their ganglia on the other, and this became customary from about the middle of the 19th century until 1884–1886. Both systems supplied "involuntary" organs and tissues. The resemblance of certain branches of the cranial nerves and of the visceral branches of the sacral nerves to the splanchnic nerves were from time to time pointed out, but there was no general grouping of the nerves supplying unstriated muscle and glands into a single system.

A partial return to an earlier nomenclature was advocated by Dastre and Morat in 1884. It was based on their observations on vaso-motor nerves. They divided motor nerves into non-ganglionated nerves subserving the "vie de relation" of Bichat, and ganglionic nerves subserving Bichat's "vie de nutrition," excluding, however, nutrition and assimilation. Under the general head of the sympathetic or ganglionic system they included certain unnamed branches of the 5th, 7th, 9th and 10th cranial nerves, so that the sympathetic system was defined in much the same way as it had been defined by Winslow.

VISCERAL NERVES. GANGLIONATED SPLANCHNIC NERVES. Gaskell (1886–1889) considered the question of the classification of nerves chiefly from a morphological standpoint. He said that certain only of the cerebro-spinal nerves had a "visceral" branch in the sense of the morphologist. These branches he placed together as visceral nerves. They arose from homologous columns of cells in the central nervous system, broken by the development of the nerves to the limbs; the vagus was homologous with the hypogastric nerve and the nervus erigens with the splanchnic nerve. Thus the visceral nerves left the central nervous system in three outflows—the cranial, the thoracic, and the sacral. This was the first definite statement that the sympathetic did not receive branches from each spinal nerve. From another point of view he divided the nerves into somatic and splanchnic, each having a ganglionated and a non-ganglionated part. The visceral nerves formed the ganglionated splanchnic part. The ganglia on the course of the visceral nerves Gaskell classified as (1) proximal or vertebral ganglia, i.e. the ganglia of the sympathetic chain from the first thoracic ganglion downwards; and (2) distal ganglia, consisting of (*a*) pre-vertebral ganglia, which included the superior cervical, the semilunar and the inferior mesenteric ganglia; and (*b*) terminal ganglia, which were scattered in or near the tissues. Gaskell considered (1886 p. 30) that the blood vessels of the mylohyoid muscle of the frog, the vessels of the limbs of mammals, and possibly other vessels, in addition to receiving nerve fibres from visceral nerves, received efferent vaso-dilator fibres which ran direct outward in the roots of the nerves.

AUTONOMIC. The observations made by Dickinson and myself (1889) on the action of nicotine gave a new method of investigating the connexion of nerve fibres with peripheral nerve cells. In describing the results of such investigations and of others suggested by it, I felt the need of a new term for the system of nerves I was dealing with. All the old terms had been used with different connotations at different times, and to use any of them with another additional connotation was but to add to the inherent difficulty of understanding the point of view of earlier writers. I called the system the autonomic nerve system (1898). It was a "local" autonomy that I had in mind. The word "autonomic" does suggest a much greater degree of independence of the central nervous system than in fact exists, except perhaps in that part which is in the walls of the alimentary canal. But it is, I think, more important that new words should be used for new ideas than that the words should be accurately descriptive. In any case the old terms have no advantage as descriptive terms.

"Vegetative" implies a contrast between plant and animal life, and to speak of a "vegetative nervous system" is to obscure one of the fundamental differences between plants and animals, viz., the absence of nerves in the one, and their presence in the other. Overlooking this we might still use "vegetative" for the nervous system in animals governing those processes in them which occur also in plants. But assimilation and metabolism are conspicuous plant processes, and control of these processes in animals is not a special function of the "vegetative" nervous system.

The fundamental drawback to the use of "involuntary" as a designation for the nervous system we are considering is that it makes subjective sensations the criterion of classification. It is inappropriate in a science based on objective observation. The term also implies that all other movements are "voluntary," whereas, in fact, all "voluntary" muscles are set in action involuntarily, and some more often than voluntarily. Further, there are a few cases of muscles which histologically belong to the class of "voluntary" muscle, but which are apparently not controlled by the will. There is no evidence that the striated muscle of the œsophagus can be contracted voluntarily; so far as it is known it is only set in action by initiating swallowing, i.e. it is set in action in exactly the same way as the "involuntary" unstriated muscle of the œsophagus. It is difficult to believe that a frog can voluntarily control its lymph hearts, or that a bird can contract its iris at will. Lastly, it is untrue that the "involuntary" actions are out of all control of the will. What is true is that the nervous mechanism set in action is different from that set in action when, for example, a limb is voluntarily moved. The will exercises more or less control over unstriated muscles and glands by recalling emotions and sensations. And apart from any special emotion, some people can by effort of will cause contraction in the involuntary unstriated muscle of the skin, and others can cause acceleration of the heart. Some physiologists have claimed that they can contract the bladder at will, and cases are recorded of voluntary inhibition of the heart, and contraction of the pupil. It can hardly be doubted that many more instances would be found if sufficient attention were directed to the subject.

The term "ganglionic" is misleading since there are ganglia on the course of all the spinal nerves and on most of the cranial nerves, and since historically the "ganglionic system" was long used in its natural signification to include these ganglia.

Sympathetic nerves have no special relation to sympathies. But the chief objection to calling the whole autonomic system sympathetic is that it confuses instead of simplifying nomenclature. Nearly all observers have used it for what was formerly called the intercostal nerve, and rarely if ever has it been taken to include the nervus erigens.

At the time I introduced the term "autonomic" there were two points of view from which the innervation of unstriated muscle and glands were regarded. From one point of view these tissues were supplied with nerve fibres partly by cerebro-spinal nerves and partly by sympathetic nerves. From the other (that of Gaskell) they were supplied by one system, which anatomically was separated into three parts by the development of the nerves to the limbs. Neither of these seemed to me properly to express the conditions. The facts that the sympathetic innervated the whole body, whilst the cranial and sacral outflows innervated parts only, and that the sympathetic had, in general, opposite functional effects from those of the other autonomic nerves, indicated that the sympathetic was a system distinct from the rest. The part of the cranial outflow supplying the eye seemed clearly to be distinct from the rest of the cranial outflow. Further, the bulbar part of the cranial outflow and the sacral outflow seemed equally clearly to form one system innervating the alimentary canal and parts developmentally connected with it. I divided (1898) the autonomic system into tectal, bulbo-sacral, and sympathetic systems, and considered that each had a different developmental history.

The theory that there is some fundamental difference between the sympathetic and the rest of the autonomic system was much strengthened by the discovery that the effects produced by adrenaline were apparently confined to effects caused by stimulating sympathetic nerves (cp. p 29). Since other drugs caused effects more or less confined to those produced by stimulating tectal and bulbo-sacral nerves, it was convenient to have a common name for these nerves, and I placed them together as the parasympathetic system (1905). The pharmacological relations of the nerve systems and some variations in the meaning attached to parasympathetic I discuss later.

I pointed out (1900) that we should expect the cells of Auerbach's and Meissner's plexuses to be on the course of the bulbar and sacral nerves, but as there was no clear proof of their central connexion, and as their obvious histological characters differed from those of any other peripheral nerve cell, I placed them in a class by themselves as the enteric nervous system. This classification is, I think, still advisable, for the central connexion of the enteric nerve cells is still uncertain, and evidence has been obtained that they have automatic and reflex functions which other peripheral nerve cells do not possess. Gaskell, in his postumous work (1916), advocated on developmental grounds the possible theory that the enteric nerve system is part of the bulbo-sacral. He did not, however, settle any of the doubtful points.

Most previous observers who had discussed the question considered that the posterior root ganglia contained some cells of the type of the sympathetic, and that the sympathetic contained cells of the type of the posterior root ganglion, some left the question open. My observations led me to believe that there was no such admixture of the different type of cells. On this theory all autonomic nerves were motor, and the afferent nerves accompanying it were not as a whole distinguishable from the other afferent nerves. How far this is true will be discussed later.

It was convenient to have some term for the general body nerves which did not belong to the autonomic system, and I used the term "somatic" since it expressed fairly well the character of the nerves.

The statement of Stricker and others that efferent vaso-dilator fibres ran direct to the limbs by way of the posterior roots was confirmed by Bayliss (1900). I had come to the conclusion that the vaso-dilator effect was produced by afferent fibres, and was due to setting free of metabolites. This theory I communicated to Bayliss, through Prof. Starling, and suggested that he should try the effect of stimulating the posterior roots after time had been allowed for degeneration. He found that the vaso-dilator effect could still be obtained, but attributed it (1901) to a direct action in the blood vessels. The nature and extent of this "antidromic" action I shall discuss later. For the present it is, I think, best classed as a property of afferent somatic nerves.

The peripheral nerves, then, excluding the olfactory, optic and auditory, may be classified as follows:

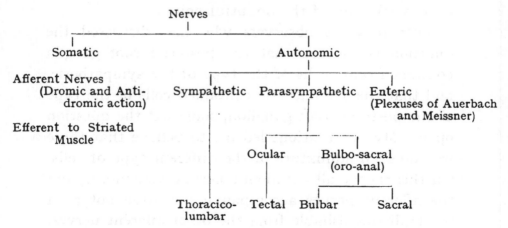

Some writers have restricted the term autonomic to

the parasympathetic system. There are other points in nomenclature which may be mentioned here.

The terms vertebral and prevertebral ganglia do not seem to me satisfactory. The cervical ganglia are prevertebral in mammals and vertebral in other vertebrates, so that the terms suggest their homology with the ganglia of the abdominal viscera in the mammal, and with the ganglia of the lower part of the chain in other vertebrates.

It is frequently necessary to describe the state of a tissue after the postganglionic fibres running to it have been cut, or the ganglion supplying it with nerve fibres has been excised, and time has been allowed for degeneration of the cut fibres. The operation is conveniently described as a *degenerative section*.

I use the terms *paralysis of a nerve*, as it is usually used, to signify that stimulation of the nerve has no effect, and not that the excitability or conductivity of the nerve has been abolished.

The terminology expressing the point of action of a chemical substance varies with different writers. The following is, I think, the most convenient. The point of action is:

the *end apparatus*, when it is not desired to specify anything further than that the action is at the periphery and can be set in action by the nerve,

the *nerve ending*, in its usual significance of an action on the terminal branches of the nerve,

the *myo-neural, or neuro-cellular junction* when it is held that there is continuity between nerve and cell, and that the action is on a partly nervous, partly muscle substance at the junction,

the *neural region* when it is held that the action is on

the cell and more or less confined to the region of the nerve ending,

the *synapse*, when it is held that there is no continuity of nerve and cell and that the action is a physical one in the neural region,

the *receptive substance* when it is held that the action is brought about by a chemical combination with a special cell constituent.

REFERENCES AND NOTES.

Winslow. Exposition Anat. de la Structure du Corps Humain. (Paris.) 1732.
Johnstone. Phil. Trans. London, 54, p. 177, 1764. Essay on the use of the Ganglions of the Nerves. (Shrewsbury.) 1771.
Bichat. Rech. Physiol. sur la vie et la Mort. (Paris.) 1800. Anat. gén. and Anat. descriptive. (Paris.) 1801.
Dastre and Morat. Rech. Exp. sur le Syst. Vaso-Moteur. (Paris.) p. 330, 1884.
Gaskell. J. Physiol. 7, p. 1, 1886; 10, p. 153, 1889.
Langley. Text book of Physiol., ed. by Schäfer, 2, p. 694, 1900. Brain, 26, p. 1, 1903. Ergeb. d. Physiol., 2. Abt. 2, p. 819, 1903.
Bayliss. J. Physiol. 25. Proc. Physiol. Soc., p. xiii., 1900. J. Physiol. 26, p. 173, 1901.
Gaskell. The Involuntary Nervous System, 1914. Published postumously. (London.) 1916.

The terms I have used were first employed as follows:
Pre- and postganglionic fibres, Proc. Roy. Soc. 52. p. 548, 1893. *Autonomic*. J. Physiol. 23, p. 241, 1898. *Axon reflex*. Soc. de Biol. T. Jubilaire, 1899, p. 220. *Bulbar and sacral nerves as one system*, Brit. Assoc. Adv. Science, Pres. Address, Physiol. Sect., 1899; *Mid-brain autonomic*. Brain, 1903. *Parasympathetic*, Address to Students of Med. and Nat. Phil. in Amsterdam, 1905; also J. Physiol. 33, p. 403, 1905; 43. p. 173, 1911.

Gaskell (1886-1889) divided the nerves, and the tissues innervated by them, into two groups, splanchnic and somatic which corresponded more or less closely with those described as arising in the embryo from a branchiomeric or dorsal segmentation and from a mesomeric or ventral segmentation respectively. This nomenclature made the masticatory, respiratory and facial muscles, splanchnic; only the limb and main trunk muscles being somatic, and it did not seem to me to be in harmony with the connotation of the terms in Physiology. In consequence, and because branchiomeric could be used when necessary for the "splanchnic" muscles, I employed somatic for all motor nerves which were not autonomic. (Brit. Assoc. Address, 1899) and later with Anderson to include also afferent fibres arising from spinal and the similar cranial ganglia. (J. Physiol. 31. p. 380, 1904.) The relation of the tissues to different segmentations was followed up by Gaskell in connexion with his theory of the origin of vertebrates. In his work published postumously (1916) his views on the phylogenetic history of the autonomic system are given. He considered that vertebrates arose from an ancestor having two serially arranged segmentations, one connected with the appendages—the splanchnic segmentation, and the other with the body—the somatic segmentation. The splanchnic segmentation gave rise to the muscles of the gut with the exception of the sphincters, and of parts developmentally connected with it. The somatic segmentation gave rise to the other unstriated muscles. Thus some striated muscles were splanchnic, and some unstriated muscles were somatic.

Both autonomic and somatic systems consist of central and peripheral parts. The central connexions of the autonomic system are imperfectly known, so that description of this system is mainly confined to a description of the preganglionic and postganglionic nerve connexions. A limitation of "autonomic" to the peripheral nerve cells was proposed by Heubner (Zntrlb. Physiol. 1912, p. 1180). Similarly Gaskell (1914) defined the involuntary nerve system as being limited to the peripheral neurons; the preganglionic fibres he called "connector." The limitation involves considerable difficulties, and Gaskell not infrequently includes the connector fibres in the involuntary system. Moreover, all nerve fibres are connector, and to distinguish the different connector fibres means much periphrasis. Lastly, it may be mentioned that Gaskell used "enteral" as synonymous with bulbo-sacral.

Voluntary production of autonomic effects. Goose skin and erection of hairs. Chalmers (J. of Physiol. 31, 1904, Proc.

Physiol. Soc., p. lx.). The action was not restricted to one region, but was produced more strongly in any desired region. Maxwell (Amer. J. Physiol. 7, p. 369, 1902), Kœnigsfeld and Zierl (Deut. A. f. klin. Med. 106, p. 442, 1912). Goose skin was produced on one side of the body on imagining this to be cold and the other side to be hot. *Acceleration of heart*, see Papers quoted by White (Heart, 6, p. 175, 1917); in White's case voluntary acceleration was also obtained after atropine. King (Bull. J. H. Hosp. 31, p. 303, 1920). For some observations on the pupil cp. Lewandowsky (Zts. f. d. ges. Neurol. 14, p. 282, 1913).

2. General Plan of Origin and of Peripheral Distribution.

At this stage it is only necessary to give the general plan of the superficial origin of the preganglionic fibres from the central nervous system. Detailed description will be given later.

The broad facts with regard to the origin of the tectal, bulbar, and sacral nerve fibres have long been known. The ciliary nerves arise from the region of the anterior corpus quadrigeminum and form part of the 3rd cranial nerve. The bulbar nerves arise from the lower part of the spinal bulb and issue in the portio intermedia of the 7th cranial nerve, and in the 9th, 10th and bulbar part of the 11th nerves. The 5th nerve has been said to send inhibitory fibres to the sphincter of the pupil and vaso-dilator fibres to the mouth, but if any effect is produced by this nerve it is probably by antidromic action. The sacral autonomic nerves arise from the middle and lower part of the sacral spinal cord and issue in sacral nerves (generally three) numbered differently in different animals.

The determination of the origin of the preganglionic fibres of the sympathetic is more recent. At the time of Gaskell's researches the sympathetic was supposed to arise from all the spinal nerves. The anatomical observations of Beck on man (1846), if they had received attention at a later time, should have raised doubt as to the origin of sympathetic fibres from the lower and upper regions of the spinal cord. He described the thoracic and lumbar nerves as having both white and grey rami communicantes, the white rami becoming smaller in passing down the lumbar region, and the cervical and sacral nerves as having grey rami communicantes only. But at this time little was known with certainty either of the origin of the "tubular" fibres which formed the chief constituent of the white rami, nor of the "gelatinous" fibres which formed the chief constituent of the grey rami. The question discussed was the degree of independence of the ganglia. Beck supported the independence theory, and the bearing of his results on the connexion of different parts of the spinal cord with sympathetic ganglia passed unnoticed. Reissner (1862) described bundles of small nerve fibres as being present in the anterior nerve roots of the dorsal region of the cord, and scattered small fibres only as being present in those of the cervical and lumbar regions, but this was not correlated with the existence of small fibres in the sympathetic.

So far as the thoracic and upper lumbar nerves were concerned the experimental evidence was ample to show that they influenced either viscera or blood vessels by way of the sympathetic. The evidence as regards the cervical, lower lumbar and sacral nerves was conflicting. If we take the number of observers who described a result as a criterion of the probability of its truth, the balance of evidence was strongly in favour of the cervical nerves sending fibres to the cervical sympathetic. Budge, François-Franck, Salkovski, Navrocki and Przybylski had all described pupillo-dilator fibres as arising from one or more of the lower cervical nerves. Cl. Bernard (1862) almost alone found no pupil effect from the cervical nerves. Dastre and Morat described the lower cervical nerves as sending vaso-constrictor fibres to the cervical sympathetic. Bever and Bezold, Boehm and Nussbaum, who investigated the origin of sympathetic cardiac accelerator fibres, considered them to arise in part from the lower cervical nerves. On the other hand the evidence was against the lower lumbar or sacral nerves sending either vaso-motor or secretory fibres to the limbs by way of the sympathetic, and the question at issue was mainly the presence or absence of vaso-motor and secretory fibres taking a direct course in the spinal nerves. The lower lumbar and sacral nerves were supposed to send fibres to the viscera by way of the sympathetic chain, but there was no satisfactory evidence for this.

Gaskell (1886) approached the subject from a histological and morphological standpoint. He found in the dog that bundles of small nerve fibres $1 \cdot 8$ to $3 \cdot 6\mu$ in diameter occurred in the anterior roots of those nerves which had white rami, but not in the anterior roots of the nerves which had grey rami, except in the case of the sacral nerves from which the nervus erigens arose, and this had no connexion with the sympathetic chain. The white rami—as had been shown by Bidder and Volkmann and others—consisted mainly of similar small medullated fibres. On tracing the grey rami towards the spinal cord he found no non-medullated fibres in the roots, inside the dura mater; the few non-medullated fibres found more peripherally he considered ran to the dura mater and thence to the blood vessels of the cord.

> A somewhat similar result and conclusion had been arrived at by Beck in man. He described a few "gelatinous" fibres in the roots, not traceable to the cord, but seeming to run to the blood vessels. (Todd and Bowman's Physiological Anat. and Physiol. of Man, 2, p. 132, 1859,

London.) The statement of Ranson that there are numerous non-medullated fibres in the posterior roots I shall discuss later.

From these and other results Gaskell came to the important conclusion that only those nerves which had white rami sent fibres to the sympathetic system. These formed his "thoracic outflow." In the dog he found white rami from the 2nd thoracic to the 2nd lumbar (the 4th lumbar of most writers). The limits of the white rami will be dealt with under the section of the sympathetic system; here it is sufficient to take them as present in the dog from the 1st thoracic to about the 4th lumbar nerve. The facts described by Gaskell showed conclusively that the very great majority of the nerve fibres passing to the sympathetic arose from a limited region of the spinal cord. They were less conclusive as to the total absence of such fibres from the upper and lower regions of the cord. As evidence of total absence Gaskell relied on two observations, first that the anterior roots of the cervical and lower lumbar nerves had no fibres less than 3.6μ in diameter. This statement is not quite accurate; there are a few fibres 2 to 3.6μ, i.e. of the size of sympathetic preganglionic fibres, in the anterior roots of the nerves having no white rami. The other point relied on was that the grey rami contained very few small medullated fibres. In fact, however, the grey rami of the cat and of some other animals contain a considerable number of small medullated fibres (cp. p. 24). Whether there are few or many it is impossible to follow them when undegenerated through the nerve trunk and into the roots in such way as to be certain that none arise from the spinal cord.

The contradiction between the conclusion drawn

from histological observations and those drawn from experiments for the most part disappeared as further experiments were made on the effect of stimulating the spinal nerves at their origin. (These experiments will be considered later under the headings of the "sympathetic" and of "antidromic action.")

There was still the possibility that the experimental method was not sufficiently delicate to detect the trifling effect which would be produced if a few fibres ran from the spinal cord to the sympathetic by way of the grey rami. Evidence on this point I obtained by the degeneration method (1896). On cutting the lumbar and sacral nerves inside the vertebral canal, either no small medullated fibres were found in the grey rami, or a few were found which there was good reason to believe arose from white rami or were afferent fibres. The function of the small nerve fibres in the anterior roots of the nerves which have no white rami remains to be determined.

The origin of the preganglionic fibres from the spinal cord may then be diagrammatically represented as in Fig. 1.

FIG. 1.   Tectal   Bulbar   Thoracico-lumbar   Sacral

The general plan of the peripheral connexion of the several parts of the autonomic nervous system is, as has been said above, that the thoracico-lumbar or sympathetic system supplies nerve fibres to all regions of the body, whilst the tectal, the bulbar and the sacral systems supply nerve fibres to special regions only.

The tectal or mid-brain system supplies the sphincter of the iris and the ciliary muscle.

The bulbar autonomic supplies almost exclusively the parts which have developed from or in connexion with the primitive alimentary canal. It innervates the mucous membrane of the nose, mouth and pharynx with the lachrymal and salivary glands, the œsophagus, stomach, liver, pancreas, the small intestine, and in some animals at any rate the anterior part of the large intestine. It innervates also the lungs (which develop as a diverticulum from the œsophagus) and the heart.

The sacral autonomic innervates the lower part of the large intestine, possibly also the upper part, the bladder, and the external generative organs (penis and vagina).

Whilst anatomically the sympathetic runs to all regions of the body, and the parasympathetic to some of these regions, it does not follow that either system of nerves has an appreciable effect on all the structures in the regions to which it runs. Thus, whilst both sympathetic and parasympathetic nerves run to the eye, it is an open question whether any part of the eye is innervated by both nerves; the bulbar system in some lower vertebrates is said not to innervate the ventricle of the heart, and there is no satisfactory evidence that the intestinal glands are innervated by either bulbar or sympathetic nerves. The details of the connexions I shall consider under the head of the separate nerve systems.

REFERENCES.

Beck. Phil. Trans. Roy. Soc. London, p. 213, 1846.
Reissner. A. f. Anat. u. Physiol., p. 125, 1862.
Claude Bernard. J. de la Physiol. 5, p. 383, 1862.
Gaskell. J. Physiol. 7, p. 1, 1886.
Langley. Ibid. 20, p. 223, 1896.

References on the origin of sympathetic fibres from the spinal cord which are quoted in the text are given in most of the larger, not too recent, text-books. References up to 1903 are given in my article in the Ergebn. d. Physiol. 2, Abth. 2, p. 820; since I shall refer to the work from 1889 onwards under the "sympathetic system," the references need not be repeated here. The opinions held in 1879–80 may be gathered from Funke and Grünhagen's Lehrbuch d. Physiol. 6te Aufg. 2, p. 713, 1879, and from Hermann's Hdb. d. Physiol. 2, Th. 1, p. 275, 1879; 4 Th. 1, p. 343, 1880.

3. The Nerve Fibres of the Autonomic System.

In the early observations which were made on the nerves it was noticed that certain parts of the intercostal nerve were reddish or greyish, whilst other parts and all the spinal nerves were white. Out of these observations arose the terms "grey and white rami communicantes" of the spinal nerves. When the microscope came into use, the white nerves were found to contain numerous fibres which were called tubular—the medullated or myelinated of to-day. They were considered to be the only form of nerve fibre, and it was noticed that in the sympathetic nerves the great majority cf tubular fibres were small. Remak (1837–8) found a different form of nerve fibre which he called "organic" on the lines of Bichat's theory. It is the non-medullated or non-myelinated fibre of to-day. He considered that it was the sole form of nerve fibre given off by the sympathetic and spinal ganglia, and that the cerebro-spinal fibres were all tubular. Bidder and Volkmann (1842), on the other hand, after a very thorough examination of a large number of nerves in different vertebrates, came to the conclusion that the sympathetic and spinal ganglia gave off most and probably all the small tubular nerve fibres of the body, that the brain and spinal cord gave off all the medium and large fibres, and that Remak's fibres were not nerve fibres. The general recognition that the "organic" fibres were nerve fibres was slow. Those who accepted it necessarily rejected Bidder and Volkmann's theory. Kölliker (1844) confirmed Bidder and Volkmann's statement that some small tubular fibres arose from the ganglia, but contested the theory that they were confined to the ganglionic system. This theory indeed received little support from anyone. The nerve fibres arising from the posterior root ganglia were eventually recognised as being almost wholly medullated, but it was a not uncommon belief that some of them were non-medullated.

Discussion of distinguishing differences between nerves running to ganglia, those arising from them, and those running direct to the tissue, died out. It is an indication of this that the question was not even mentioned by Max Schulze in his account of nerve fibres in Stricker's Histology in 1870.

After a long interval a return to generalisation was made by Gaskell (1886). He put forward the theory that all the efferent "visceral" nerves passing from the central nervous system were small medullated fibres, that they lost their medullary sheath in the ganglia, and that the ganglia gave off non-medullated fibres. The simplicity of this theory and its obvious truth in the case of many nerve fibres in the mammal led to its obtaining a considerable degree of acceptance. The theory, it will be seen, consists of three parts. The first, viz., that all efferent "visceral" fibres as they leave the brain or cord, i.e. all pre-ganglionic fibres are small medullated fibres has been confirmed or not contested by subsequent observers. The second and third parts of the theory, viz., that the pre-ganglionic fibres lose their medulla in the ganglia, and that the ganglia give off non-medullated fibres, and these only, was important, since, if true, a means was afforded of allowing preganglionic to be distinguished from postganglionic fibres and of determining whether a ganglion had sympathetic nerve cells in it. Unfortunately the matter is less simple.

In the rabbit, as I pointed out in 1892, the great majority of the preganglionic fibres which run to the sacral and coccygeal ganglia lose their medulla a considerable distance before they reach the ganglia in which they end. In all mammals a loss of medulla appears to occur in some of the corresponding fibres, and in the majority of the preganglionic fibres of the splanchnic and vagus nerves. As to the last part of the theory, viz., that the ganglia do not give off medullated fibres, Gaskell a little later (1889) noticed, as had been noticed by Bidder and Volkmann, that the short ciliary nerves were medullated, so that the theory did not hold for the tectal autonomic. Nor does it hold for the sympathetic nerves of the dog and cat. In the cat I found (1896) the 7th lumbar grey ramus to have more than 300 medullated fibres, and the anterior branches of the superior cervical ganglion to have several hundreds, and showed by the degeneration method that they were nearly all postganglionic. Billingsley and Ranson (1918) counted in different cats about 1500 to 4000 small medullated fibres in the grey rami given off by this ganglion. In the grey rami of the dog the medullated fibres are fewer; in several thoracic rami I found 10 to 25 only, but there were many in the lumbar rami. In the rabbit the medullated fibres of the grey rami are few, and possibly in this animal none of them are postganglionic; but as the preganglionic are apt to become non-medullated, the two classes of fibres cannot be distinguished with certainty. The observations of

Bidder and Volkmann showed that in the frog, and probably in fish, reptiles and birds the ganglia gave off numerous small medullated fibres. In fact, in vertebrates other than mammals, the rami communicantes of the spinal nerves consist wholly, or almost wholly, of medullated fibres; grey rami communicantes are not present. On the other hand, most of the postganglionic fibres passing to the abdominal viscera in lower vertebrates are non-medullated.

The medullated postganglionic fibres of the mammal as a rule, lose their medulla at some point on their course to the periphery, and I am not certain that any remain medullated up to their final branches.

Sherrington (J. Physiol. 17, p. 248, 1894) did not find any medullated degenerated fibres in the muscular and cutaneous nerves of the cat after excision of the lumbar and upper sacral ganglia in the cat. Some of the branches of the sciatic in the frog contain a bundle of non-medullated fibres, with a few small medullated in them, though the postganglionic fibres entering the sciatic are medullated.

No difference in function has been found between medullated and non-medullated postganglionic fibres. The hypothesis I would suggest as to the proximate cause of the existence of the two kinds of nerve fibres is that cells with non-medullated fibres were the first in phylogeny to migrate from the central nervous system, a later migration occurring when a further specialisation of the central nervous cells had occurred, and that the cells of this migration gave rise to medullated fibres. On this hypothesis, the two forms of embryonic cells have persisted to a varying degree in different vertebrates, each form giving rise to its own kind of axon.

Gaskell gives $1 \cdot 8$ to $3 \cdot 6 \mu$ as the size of the small fibres in osmic acid preparations. Some preganglionic fibres are a trifle larger than this—up to 4μ in diameter. The sympathetic postganglionic fibres are rarely larger than $2\cdot 5\mu$; the amount of myelin in them varies; in the mammal, at any rate, it may form a very thin layer requiring deep staining to make it obvious. In transverse sections of grey rami stained with osmic acid, fibres with very thin sheath may easily be overlooked.

> There is much that is still unknown with regard to the sheath of the non-medullated fibres. In the sacral region the preganglionic nerve fibres which have lost their myelin are not obviously different from postganglionic fibres which have none. Yet the sheath of a postganglionic fibre has little resemblance to the sheath (neurilemma) of a medullated fibre. Tuckett describes the postganglionic fibres near their origin as having a relatively thick sheath which readily breaks up into fibrillæ, in which the nuclei are imbedded. But at the periphery the non-medullated fibres forming the terminal plexus appear to be without sheath. Where the sheath is lost, if it is lost, is unknown. The sheath of the non-medullated fibres of the thoracic vagus seems to me to be different from that of the sympathetic fibres.

After section of postganglionic nerves the nerves cease to be irritable in a few days, but when teased out in osmic acid preparations the non-medullated fibres may maintain their usual appearance for two or more weeks. Tuckett (1896) found that the central axon core ceased to stain normally with methylene blue about the time the irritability disappeared. Recently non-medullated fibres have been described by Ranson (1912) and by Boeke and Dusser de Barenne (1919) as degenerating very much later than medullated fibres. They treated the nerves by Bielschowsky's silver method, and I consider that what they took for undergenerated non-medullated fibres was only the sheath of the fibres.

In the preceding account I have dealt with certain points common to all autonomic nerve fibres. Some further account of the histological characters of the grey and white rami will be given under the sympathetic system.

REFERENCES AND NOTES.

Remak. Froriep's Notizen, 1837. Observationes anat. et microc. de systematis nervosi structura (Berol), 1838.
Bidder and Volkmann. Die Selbständigkeit d. Symp. Nervensystems (Leipzig), 1842.
Kolliker. Die Selbständigkeit and Abhängigkeit d. Symp. Nervensystems (Zurich), 1844.
Gaskell. J. Physiol. 7, p. 1, 1886. Ibid. 10, p. 153, 1889.
Langley. Phil. Trans. London, 183, B. p. 85, 1892.
Tuckett. J. Physiol. 19, p. 267, 1896.
Langley. Ibid. 20, p. 55, 1896.
Ranson. J. Comp. Neur. 22, p. 487, 1912.
Billingsley and Ranson. Ibid. 29, p. 367, 1918.
Boeke and Dusser de Barenne. Proc. k. Akad. Amsterdam, 21, 1919.

Numerous small nerve fibres are present in the posterior roots of all spinal nerves; whether these are especially concerned with antidromic action is unknown.

4. The Specific Action of Drugs on the Sympathetic and Parasympathetic Systems.[1]

One reason for dividing the autonomic nerves into sympathetic and parasympathetic is, as I have said, that the sympathetic nerves supply all parts of the body, whilst the remaining nerves supply special parts only. And this division is convenient in describing pharmacological results. Adrenaline and certain related substances produce effects which are in nearly all cases like those produced by stimulating sympathetic nerves. Pilocarpine, muscarine and arecoline, on the other hand, whilst having some differences in action, produce effects which are in most cases like those produced by stimulating parasympathetic nerves. Most of the effects produced by physostigmine and by choline are such as would be produced by increasing the excitability or by stimulating parasympathetic nerves. Atropine abolishes the effects of the pilocarpine group of drugs, dulls, and in some cases abolishes the effect of stimulating the parasympathetic nerves. The facts show that there is a close relation between the action of the drugs and innervation by sympathetic and parasympathetic nerves respectively, and they suggest that there is a fundamental difference between the two systems. The facts are also of importance with regard to the

[1] I discuss this subject before that of the tissues innervated by the sympathetic chiefly because the action of certain drugs is commonly taken as showing whether a tissue is innervated by bulbo-sacral or sympathetic nerves, and to give the evidence here will save repetition.

action of hormones on the tissues. It is generally held that the adrenaline formed in the body sustains sympathetic action; and it has been held that choline, which is also formed in the body, sustains parasympathetic action.

Taking adrenaline and pilocarpine as respectively representing the two classes of drugs, we may consider the cases in which there is, or appears to be, a lack of correspondence between the action of the drugs and that of nerve stimulation.

> Barger and Dale (J. Physiol. 41, p. 19, 1910) use the term sympathomimetic to describe the action of amines. This term, and parasympathomimetic may conveniently be used to imply that the effects produced by the drugs resemble broadly, <u>but not exactly</u>, the effects produced by nerve stimulation.

Adrenaline and sympathetic effects.

The only structures markedly influenced by sympathetic nerve stimulation which are not influenced by adrenaline after nerve section are the sweat glands of the cat and of some other mammals. Dieden (1916) observed that injection of adrenaline into the subcutaneous tissue of the cat's foot in certain conditions caused secretion, but this, I find (1921), is due to the fluid in which it is dissolved, and not to the adrenaline itself. A doubtful difference in action is presented by the statements regarding the tone of striated muscle. On the basis of the effects of nerve section, the sympathetic is held by some observers (cp. p. 71) to cause tonic contraction. This has not been found to be produced by adrenaline, but it has also not been found to be caused by sympathetic nerve stimulation.

When the relative effect of sympathetic stimulation and of adrenaline on different tissues is compared, there are some striking instances of lack of correspondence. Thus weak stimulation of the sympathetic in the cat causes contraction of the cutaneous arteries, contraction of the tunica dartos of the scrotum and of the erector muscles of the hairs. A very small amount of adrenaline causes contraction of the cutaneous arteries, a larger amount is required to cause contraction of the tunica dartos, and a very much larger amount is required to cause contraction of the erector muscles of the hairs.

> There is a similar difference in the sympathetic and the adrenaline effects on the arteries and on the unstriated muscles causing movement of the feathers of the fowl (Langley, J. Physiol. 30, p. 240, 1903). The relatively slight effect on skin muscles is not, however, universal, for adrenaline readily causes erection of the quills of the hedgehog (Lewandowsky, 1900), of the hairs of the tail of the marmot (Kahn, A. (Anat. u.) Physiol., p. 239, 1903), of the hairs of the mongoose, and goose skin in man (Elliott, 1905). Elliott connected the degree of action of adrenaline with the normal frequency of sympathetic stimulation.

There is also some lack of correspondence in readiness of production of contraction and inhibition by sympathetic nerves and by adrenaline when the sympathetic has both contractor and inhibitory fibres. The most striking instance is that of the bucco-facial region of dog, in which adrenaline has not been found to cause any flushing but only contraction. In the cat's bladder on the other hand, adrenaline causes inhibition more readily than does the sympathetic.

> In the bladder of the cat, Langley (1901) and Elliott (1905) found no preliminary rise of internal pressure with adrenaline such as is caused by sympathetic stimulation. Edmunds and Roth (J. Pharm. exp. Ther. 15, p. 189, 1920) usually obtained a preliminary rise, but on the excised trigonal region—the part which contracts on sympathetic stimulation—they frequently failed to get contraction.

Adler (A. exp. Path. Pharm. 83, p. 248, 1918) usually obtained inhibition of the excised frog's bladder with adrenaline; the ordinary effect of sympathetic stimulation is contraction. The different effects may be due to differences of dose, but this has not been shown.

Further, some effects have been obtained with adrenaline which either have not been obtained by sympathetic stimulation or which have not been obtained with certainty. More decided effects on cerebral vessels have been obtained by adrenaline than by sympathetic stimulation. Except in the nictitating membrane of the frog, no certain effect of the sympathetic on capillaries has been described, but there is some evidence that adrenaline may cause either contraction or dilatation of capillaries. Capillary dilatation may be the cause of the fall of blood pressure which a small amount of adrenaline produces in carnivora; the fall has, however, been usually attributed to an action on arteries, and this also has not been obtained in normal conditions by sympathetic stimulation. Adrenaline has also been said to cause a slight movement of pigment granules in the deep-lying pigment cells of the frog, an effect not produced, so far as is known, by sympathetic nerve stimulation (cp. p. 58).

Fall of blood pressure. Dilatation of blood vessels. Moore and Purinton found that suprarenal extract in minimal amount caused a fall of blood pressure in dogs (A. ges. Physiol. 81, p. 483, 1900), and Hoskins and McClure showed that the effect was due to adrenaline (A. Int. Med. 10, p. 353, 1912), see also Cannon and Lyman (Amer. J. Physiol. 31, p. 376, 1913). The amount required to cause a fall of blood pressure varies in different tissues (Hartman, ibid. 38, p. 438, 1915, and other papers in Vols. 43 and 44). Adrenaline has not been found to cause fall of pressure in the rabbit, but Ogawa (A. exp. Path. u. Pharm. 67, p. 89, 1912) states that perfusion of the intestine, kidney and skin vessels of the rabbit with very dilute adrenaline causes dilatation when the perfusion is kept up for some time (cp. also Bauer and Fröhlich A. exp. Path. u. Pharm. 84, p. 33, 1908). Desbouis and Langlois state that in minimal dose its effect on the vessels of the lung is dilatation (C.R. Soc. Biol. 72, p. 674, 1912), and Rothlin (Habilitationsschrift Zürich, 1920) finds a similar effect on the isolated renal artery.

Cerebral vessels. Biedl and Reiner (A. ges. Physiol. 79, p. 193, 1900) and Wieschovski (A. exp. Path. u. Pharm. 52, p. 389, 1905) came to the conclusion that adrenaline caused strong contraction of the cerebral vessels. Dixon and Halliburton, in perfusion experiments, found slight dilatation only which they thought was confined to the larger arteries (Quar. J. exp. Physiol. 3, p. 315, 1910), Wiggers obtained contraction (Amer. J. Physiol. 14, p. 452, 1905; J. Physiol. 48, p. 109, 1914). Cow in ring preparations of the cerebral arteries found slight dilatation (J. Physiol. 42, p. 125, 1911).

Veins. According to Gunn and Chavasse, adrenaline causes rhythmic contraction in ring preparations of the superior vena cava and contraction of the jugular veins of the sheep (Proc. Roy. Soc. B. 86, p. 192, 1913). Donegan states that in the cat adrenaline in perfusion experiments has no effect on the large veins near the heart (J. Physiol. 55, p. 226, 1921).

Capillaries. On local application of adrenaline to the intestine I found intense pallor, and suggested that it was due to contraction of the capillaries (J. Physiol. 27, p. 248, 1901). Protopopov (Abstract in J. de Physiol. Path. gén., p. 1122, 1904), on injecting adrenaline, observed pallor of the brain, which he thought was probably due to a direct action on the capillaries since the visible arteries did not contract. Cotton, Slade and Lewis obtained pallor in the forearm on local injection of adrenaline in man after stopping the blood flow by compression of the upper arm (Heart 6, p. 246, 1917). Dale and Richards describe adrenaline as causing marked dilatation of capillaries in carnivora (J. Physiol. 52, p. 110, 1919). Krogh obtained dilatation of capillaries in the tongue of the frog on local application of adrenaline (J. Physiol. 53, p. 408, 1920). Elliott, on local application of adrenaline to the capillaries of the nictitating membrane of the frog, found no effect (J. Physiol. 32, p. 414, 1905); these are the only capillaries which have been found to contract on sympathetic stimulation (cp. p. 64). The vaso-dilation caused by histamine, which Dale and Richards attribute to capillary action, is said by Schenk (A. exp. Path. u. Pharm. 89, p. 332, 1921) to be prevented by adrenaline.

Adrenaline has been described as having on striated muscle several effects which have not been described as occurring on sympathetic nerve stimulation. All recent evidence tends to show that adrenaline in the amount at all likely to be present in the blood has no effect on the height or duration of the single muscular contraction, but it tends to show that in some amount beyond this it has an action similar to that of fatigue. How far this action is due to the substances which are mixed with adrenaline in the preparations used has not been clearly settled. There is, however, much evidence that when a muscle has been fatigued, a partial restoration is brought about by dilute solutions of adrenaline (Dessy and Grandis and others), and there is some evidence that adrenaline partially antagonises the action of curari. The effect on fatigue has been attributed by most observers to an action of adrenaline throughout the muscle fibres, and not to an action confined to the myo-neural junction, or to the neural region of the muscle fibres. Other instances of apparent lack of correspondence between the effects of adrenaline and of sympathetic stimulation are mentioned below under "reversed action" (cp. p. 37).

Action on muscle. Oliver and Schafer (1895) described an increase in height and duration of the single contraction of muscle (frog and dog) as being caused by suprarenal extracts. Boruttau (1899) found both by indirect stimulation and by direct stimulation of the curarised frog's gastrocnemius muscle that adrenaline caused effects similar to those caused by fatigue. Panella (A. Ital. de Biol. 48, p. 430, 1907) obtained no effect on the height of contraction of excised frog's muscles, or on normal mammalian muscle. Kuno found (1915) that ·006 p.c. adrenaline had no effect on the curve of single muscle contraction of the frog's sartorius muscle. An absence of effect was also noted by Takayasu (Quar. J. exp. Physiol. 9, p. 347, 1916) with ·0001 p.c.

adrenaline. Schafer (Endocrine Organs, p. 65, 1916) attributed his earlier results to substances other than adrenaline in the extracts. Takayasu described a lessened height and duration of contraction on immersing the frog's sartorius in adrenaline ·0002 p.c. and upwards. Okushima (Acta Schola med. Kioto 3, p. 251, 1919), on indirect stimulation of the gastrocnemius of the frog, obtained increased height of contraction with adrenaline ·0001 p.c., and nerve paralysis with ·01 p.c. (The Parke, Davis preparation used contains chloretone, and the effect of this was not tested). On direct stimulation of the muscle he found only a slight decrease with ·01 p.c. adrenaline.

Dessy and Grandis (A. Ital. de Biol. 41, p. 225, 1904) found in a frog-toad (Leptodactylus) that injection of adrenaline partially restored the height of muscular contraction which had been reduced by a series of single induction shocks. A similar but less effect was found on adding adrenaline to excised muscles. Panella (Ibid. 48, p. 430, 1907), in frogs, toads, rabbits and guinea pigs, confirmed an effect of adrenaline on fatigued muscles. Cannon and Nice (Amer. J. Physiol. 32, p. 44, 1913) showed that the effects previously produced were largely due to circulatory changes, but gave additional evidences of a direct action of adrenaline on muscular contraction. The most complete experiments on the mammal are those of Gruber (Ibid. 33, p. 358; 34, p. 89, 1914). Gruber and Fellows (Ibid. 46, p. 472, 1918) and on amphibia by Guglielmetti (Quar. J. exp. Physiol. 12, p. 139, 1919). Other references are given in these papers. A recovery from fatigue produced by direct stimulation of the muscle has been found on giving adrenaline by Gruber in the curarised cat's muscle and by Guglielmetti in the frog's gastrocnemius. Okushima did not find that adrenaline had an effect on the fatigue of the frog's gastrocnemius.

The experiments so far made do not definitely decide whether the action of adrenaline on fatigue is due to an action throughout the muscle or not. In many of the experiments the nerve fibres may have been stimulated. Boruttau and Gruber found an effect after curari, but it is possible that the amount of curari did not reduce the excitability of the neural region of the muscle to that of the non-neural region.

Panella (A. Ital. de Biol. 47, p. 17, 1907) and Gruber (op. cit. vol. 34) describe adrenaline as partially antagonising the paralysing action of curari.

Some other observations on the effect of adrenaline on striated muscle I give on p. 76.

Lastly, one action of adrenaline has been attributed to stimulation of the end apparatus of parasympathetic nerves. Luckhardt and Carlson (1921) find that whilst adrenaline causes contraction of the pulmonary arteries of the frog and turtle, sympathetic stimulation does not. They find that in these animals the vaso-constrictor fibres for the lung run in the vagus and are paralysed by atropine.

The account given above may be summarised as follows:—There is only one clear case in which sympathetic stimulation has a marked effect and adrenaline has none (the sweat glands), but the degree of effect of the two forms of stimulation do not run parallel in the several tissues, the balance of contractor and inhibitory effect is not the same, and one or other effect caused by sympathetic stimulation may apparently be absent on injecting adrenaline. Conversely adrenaline has an effect in some cases in which stimulation of the sympathetic has not been shown to have any; in general these are cases in which very dilute adrenaline is used. In one case (lung vessels of frog and turtle) adrenaline is said to have the same effect as parasympathetic nerve stimulation.

Pilocarpine and parasympathetic effects.

Nearly all the effects which are caused by stimulation of parasympathetic nerves are also caused by pilocarpine. The most marked exception is that the sacral autonomic nerves cause inhibition of the retractor penis, and pilocarpine causes contraction. The degree of effect produced by the two forms of stimulation is not always the same; it differs strikingly in the small intestine. The action of pilocarpine is not confined to the tissues innervated by parasympathetic nerves, it causes secretion from sweat glands, contraction of the uterus, and contraction or dilatation of arterioles. It causes also contraction of the spleen, an effect which has not been found to be produced by stimulation of the vagus. Similarly, the paralysing action of atropine is not confined to parasympathetic nerves, it paralyses the sympathetic secretory fibres of the submaxillary gland, and of the sweat glands of the cat, the secretory nerves of the skin glands of the frog, and, according to Mislavski and Borman, the sympathetic secretory fibres of the prostate gland.

Uterus. Cushny (J. Physiol. 41, p. 238, 1910) found that in the cat injection of pilocarpine caused an effect similar to that caused by sympathetic stimulation, i.e. as a rule inhibition in the non-pregnant animal and contraction in the pregnant animal. Dale and Laidlaw (Ibid. 45, p. 1, 1912) found the inhibition to be inconstant and not to occur in the excised uterus; they attributed the inhibition in the body partly to a liberation of adrenaline and partly to stimulation of sympathetic ganglia by pilocarpine. The uterus of the guinea-pig, whether pregnant or non-pregnant, they found to be contracted only by pilocarpine. Gunn and Gunn, however (J. Pharm. exp. Ther. 5, p. 527, 1914) sometimes obtained inhibition in the excised uterus of the guinea-pig, and in two experiments inhibition in that of the rat, though more often there was no decided effect. Dale and Laidlaw did not find the contractor action of pilocarpine to be abolished by ergotoxine. According to Gōhara (Acta Sch. Med. Kioto. 3, p. 363, 1920) atropine abolishes the contractor effect of adrenaline on the uterus of the rabbit.

For contraction of the retractor penis by pilocarpine cp. Bottazzi (A. Ital. de Biol. 65, p. 265, 1916). Edmunds (J. Pharm. exp. Ther. 15, p. 201, 1920) found that pilocarpine still caused contraction after ergotoxine had abolished the motor effect of adrenaline. Contraction of the spleen is mentioned by Harvey (J. Physiol. 35, p. 116, 1906) and by Dixon (Manual of Pharm., p. 88, 1919).

Blood Vessels. Pilocarpine causes flushing of the skin of man. And experiments on animals indicate that on injection it has a peripheral vaso-dilator action. (Langley,

J. Anat. and Physiol. 10, p. 66, 1875; Reid Hunt, Amer. J. Physiol. 44, p. 254, 1918.) According to Brodie and Dixon (J. Physiol. 30, p. 485, 1904), pilocarpine causes contraction of vessels in the intestine and limbs in perfusion experiments. It has been said to cause slight dilatation of lung vessels (Brodie and Dixon, op. cit. supra., p. 488; Baehr and Pick, A. exp. Path. u. Pharm. 74, p. 55, 1913), and also contraction (Beresin, A. ges. Physiol. 158, p. 219, 1914). Dixon and Halliburton (Quar. J. exp. Physiol. 3, p. 315, 1910) found that it caused dilatation of the vessels of the brain. Amsler and Pick did not find that it had any effect on the perfused vessels of the frog (A. ges. Physiol. 85, p. 61, 1919). Dixon (Manual Pharmacol., 4th edit., p. 86, 1919) states that it stimulates the end apparatus of the accelerator nerve. There are other instances of a difference of effect in pilocarpine and parasympathetic nerve stimulation.

Paralysis of sympathetic fibres by atropine. Langley, Submaxillary gland of the cat (J. Physiol. 1, p. 97, 1878; Stricker and Spina, Skin of glands of the frog (Med. Jahrb., p. 355, 1880); Mislavski and Borman, Prostate gland (Centrlb. Physiol. 12, p. 181, 1898).

Reversal of effect of sympathetic and parasympathetic drugs.

In the preceding account I have mentioned some cases in which a very small amount of adrenaline produces effects different from those caused by larger amounts and effects which have not been produced, or only doubtfully, by nerve stimulation. These are often spoken of as a reversal of adrenaline action so as not to prejudge the question whether they are normally produced in the body by nervous action or not. In perfusion experiments on the frog's heart and blood vessels, reversed action has been obtained by varying the amount of Ca salts in the perfusion fluid, by making the Ringer's fluid slightly acid, by previous treatment with another drug and by prolonged action of adrenaline. Whilst in some cases the reversal has been taken as showing the presence of a small number of nerve fibres antagonistic in action to that of the majority, in others the reversal has been taken to show that the action is determined by the condition of the tissue and is not an expression of normal nerve action, and in others again it has been taken as showing an action of a "sympathetic drug" on parasympathetic end apparatus, or of a "parasympathetic drug" on sympathetic end apparatus. One difficulty in interpreting the results lies in determining how far they are due simply to the displacement of one adsorbed substance by another. As they concern us here, they give examples in which a relation of drug action to nerve systems has not been shown.

The reversed action of adrenaline after ergotoxine, since it is commonly held to be due to stimulation of sympathetic inhibitory nerves, will be considered under the head of the sympathetic system.

Pearce (Zts. f. Biol. 62, p. 243, 1913) found in perfusion experiments on the vessels of the frog 4, 6, and 29 days after the sciatic had been cut, and also after perfusion with pure NaCl, that a small amount of adrenaline caused vaso-dilation provided curari had been given, and occasionally without it. Diastolic cessation of the heart beat of the frog by adrenaline was obtained by Burridge after perfusion with 3 p.c. sodium phosphate (Quar. J. exp. Physiol. 5, p. 368, 1912) and by Kolm and Pick after decrease of calcium in the perfusion fluid (A. ges. Physiol. 184, p. 79, 1920). Kolm and Pick also found that adrenaline stopped the heart in diastole when given after a certain amount of acetylcholine; they attributed the effect to stimulation of the end apparatus of vagus fibres since it was prevented by atropine. Snyder and Campbell (Amer. J. Physiol. 47, p. 199, 1920) observed change of the action of very dilute adrenaline from vaso-constriction to vaso-dilation when the fluid was made slightly acid.

Kato and Watanabe after repeated injection of adrenaline in the dog, found that adrenaline placed on the conjunctiva caused contraction of the pupil (Tokio J. exp. Med. 1, p. 73,

1920). This, however, may possibly have been due to an increased responsiveness of the blood vessels; it was found by myself and by Elliott that dilatation of the pupil in the dog requires a relatively large dose of adrenaline. After perfusing the frog's heart with Ringer's fluid containing adrenaline and excess of Ca, Pick found that acetyl-choline—a parasympathetic drug—caused contracture instead of inhibition; the effect was not prevented by atropine, but was prevented by ergotoxine, and Pick considered it to be due to an action of acetyl-choline on the end apparatus of the sympathetic nerves. (Wien. klin. Wochens, No. 50, 1920.)

Theories on the relation of drugs to nerve systems.

The relation of drugs to particular nerve systems is not confined to those mentioned above. Thus curari paralyses efferent nerves leaving the spinal cord; it paralyses first somatic motor nerves, the phrenic being paralysed somewhat later than the nerves supplying the skeletal muscles, it then paralyses preganglionic fibres, and in mammals it has little or no paralysing action on postganglionic fibres. In each case it is the end apparatus of the nerve fibres which is paralysed. The action of nicotine in mammals is also in general restricted to the connexions of the nerves leaving the spinal cord. It first stimulates the end apparatus of nearly all preganglionic nerves (or the nerve cells direct) and in some animals paralyses them, it then paralyses somatic motor nerves. A more limited relation of a drug to nerve systems was found by Dale (1906). Ergotoxin paralyses, or nearly paralyses, motor contractor sympathetic nerves, whilst leaving unaffected inhibitory sympathetic nerves and all parasympathetic nerves.

In no case, however, is there complete correspondence between the actions of a drug and the effects of nerve stimulation. And the action of some drugs seems to have little relation to nerve systems. Thus

Dixon (1903), though he found that apocodeine, as nicotine, paralysed all efferent nerve fibres leaving the central nervous system, found also that it paralysed postganglionic fibres without relation to any nerve system. A large dose of apocodeine paralysed both sympathetic and parasympathetic nerves supplying the iris and the heart, prevented both adrenaline and pilocarpine from affecting the intestine, and prevented adrenaline from causing a rise of blood pressure, in each case leaving barium chloride with its ordinary action. On the other hand, apocodeine did not prevent either adrenaline or pilocarpine from having an action on the bladder. The action thus seems to depend rather on the nature of the tissue than on the nervous system supplying it.

Whilst I confine the following discussion to the problem of the cause of the correlation between adrenaline and sympathetic nerve action on the one hand, and that of pilocarpine and the parasympathetic nerve action on the other, I consider that the general statements are applicable to all cases of specific drug action.

It has been repeatedly shown that adrenaline effects continue after degenerative section of postganglionic nerve fibres (Lewandowsky, Langley, Elliott), and it has been shown that the nerve endings in the arteries disappear in consequence of such section. (Eugling, Langley.) It may then be taken as certain that adrenaline acts peripherally of the nerve endings, that is peripherally of the terminal nerve branches. The evidence that pilocarpine also acts peripherally of the nerve endings rest on a narrower basis, and at the moment depends chiefly on the observation of Anderson that pilocarpine causes contraction of the pupil after degenerative section of the short ciliary nerves.

Anderson and myself (J. Physiol. 31, p. 423, 1904) found that pilocarpine on local injection caused secretion of sweat in the foot of the cat six weeks after section of the sciatic, though less secretion than on the intact side. Anderson (Ibid. 33, p. 414, 1905) found the effect on the pupil mentioned above, but also found that the effect of physostigmine disappeared a few days after the nerve section. Horsley (System of Med. by Allbutt and Rolleston 7, p. 579, 1910) observed in most cases of spinal compression in man that subcutaneous injection of pilocarpine caused secretion of sweat in regions supplied by nerves above the compression, but not in the lower regions. In these cases there is no reason to suppose that the nerve endings had degenerated, so that the decrease of irritability would appear to be due to the absence of normal nerve impulses as in the salivary glands after section of the chorda tympani. Burn (unpublished observations), after degenerative section of the sciatic of the cat on one side, found that injection of pilocarpine in regions other than the foot caused no secretion in the foot. The conclusion to be drawn from the secretion caused locally by local injection is rendered doubtful by the fact I have mentioned above that local injection of fluid may cause secretion in the normal animal. In recent experiments I have found a great decrease of excitability to pilocarpine a fortnight after section of the sciatic; the comparative effects of fluid alone and of pilocarpine solution I have not yet determined.

Much of the selective action of chemical substances is *prima facie* accounted for by the fact that the chemical characters of the cells of any one tissue are on the whole more alike than are those of different tissues, and this holds also for their physical characters. Similarly some difference in the selective action on the cells of any one tissue would be expected since their characters vary, thus in glandular tissues the chemical and physical characters of a liver cell are not the same as those of a pancreatic gland cell. In other words it seems inevitable that the action of chemical substances should vary in accordance with the degree of differentiation of the different cells, and, so far as the different groups of cells are innervated by different nerve systems, the action of chemical substances would then be correlated with different nerve systems.

But this explanation of specific drug action is insufficient in itself when the same tissue is innervated by the two nerve systems, and a drug produces the effect of one and not of the other. *A priori* the result might be due either to the nervous systems being different and producing a definite and different change in the cells in which they end, independently of the nature of the cells, or to a differentiation of cell substance independently of the nature of the nerve systems.

So far as the action of adrenaline on unstriated muscle is concerned the former theory has been taken by Elliott.

Brodie and Dixon (1904), in discussing the point of action of adrenaline and of some other drugs, had taken the nerve and muscle to be continuous, and the substance at the junction—the myo-neural junction—to be partly nervous, but under the trophic influence of the muscle as well as of the nerve. This theory altered to the least possible extent the older theory that drugs acted on the terminal branches of the nerve—a theory I had contested in the case of nicotine. Elliott took a similar view with regard to the action of adrenaline on unstriated muscle, and considered that if a sympathetic nerve joined an unstriated muscle cell it produced at the myo-neural junction a substance responsive to adrenaline without reference to the inherent properties of the muscle. If in accordance with present evidence we take the nerve impulses to be the same in sympathetic as in parasympathetic nerves, the change produced by the sympathetic must, on this theory, be due to its special chemical characters.

> Later (1907) Elliott considered that the action of nerve on cell could be equally well explained as the synaptic membrane theory, i.e. with discontinuity of nerve and cell. This makes the chemical action required much less likely, for it involves the secretion of a substance from the nerve endings.

The theory, however, breaks down as a general explanation of the relation of adrenaline to the sympathetic and of pilocarpine to the parasympathetic system, for it requires that adrenaline should cause secretion of sweat, and it does not, and that pilocarpine should not cause secretion of sweat or contraction of the uterus, and it causes both.

The theory I have put forward is the latter of those mentioned above, viz., that the effects of the drugs are due to cell differentiation in phylogenetic development independently of the nature of the nerve. I take it that at an early stage, the organism consisted of the elements of the tectal and bulbo-sacral systems and cells connected with them, and the elements of the epidermis. Since these developed under a common chemical and physical environment, they had certain characters in common which enabled them to be acted upon to a varying extent by parasympatho-mimetic drugs. At a later stage the sympathetic nerve system developed, and the cells with which they became connected had then for the most part acquired other chemical and physical characters (probably in part in consequence of the presence of adrenaline), which enabled them to be acted on by sympatho-mimetic drugs. The effect of the sympathetic nerves depended on how far the cells with which they became connected had acquired the newer characters, and on how far they retained the earlier ones. Thus the sweat glands of the mammal, in consequence of the special chemical character of the epidermic cells from which they developed, did not become responsive to adrenaline, and retained the early character which enabled them to respond to pilocarpine. Since they developed late they became connected with the sympathetic system. Among environment conditions are to be reckoned nerve stimuli. These produced a change which was at its maximum at the point where the impulses entered a cell and extended a variable distance beyond it, the whole forming the neural region of the cell. The change, which sometimes increased the responsiveness of the cell to certain drugs, and sometimes gave rise to it, was dependent on the nature of cell, and would have been produced by stimuli proceeding from any nerve. The known physical characters of drugs are insufficient to account for the effects they produce, though they account for a difference in rate of action; in consequence I consider that there is a chemical combination between the drug and a constituent of the cell—the receptive substance. On the theory of chemical combination it seems necessarily to follow that there are two broad classes of receptive substances; those which give rise to contraction, and those which give rise to inhibition. I assume that both parasympathetic and sympathetic nerves can become either contractor or inhibitor, and that the effect produced by stimulating them depends upon the relative amount of contractor and inhibitor receptive substance connected with them in the cells. In general, the sympathetic nerves became contractor or inhibitor as the bulbo-sacral nerves were inhibitor or contractor. The minor differences mentioned earlier between the degree and nature of the effect produced by sympathetic nerve stimulation and adrenaline I attribute to the nerve impulses affecting receptive substances not affected by adrenaline. As a pure hypothesis I have suggested that the receptive substance is—on the lines of Ehrlich's theory—a side chain of the molecule of the general cell substance, and that it is in looser combination in the neural than in the non-neural region.

On this theory it is intelligible that pilocarpine should act not only on all tissues innervated by parasympathetic nerves, but also on some tissues which are innervated by the sympathetic only. And it is not inconsistent with the possibility that adrenaline may act on some tissues which are not innervated by the sympathetic, or that acetyl-choline and histamine may cause vaso-dilation by acting on a receptive substance in capillaries receiving no efferent nerve fibres.

The present state of knowledge does not seem to me to allow sure conclusions to be drawn as to the course of differentiation in tissues which have a double innervation. The chief difficulty is that it is not known whether the cells of these tissues are of two kinds, one innervated by the parasympathetic system and the other by the sympathetic, or whether each cell receives nerve fibres from both systems. The hypothesis that the nerve fibres of the two systems end in separate cells, is, as I have said earlier (1905), that which is easier to bring into harmony with the general theory given above. It is *a priori* probable that a certain degree of increased complexity of chemical constitution should cause cell division in which one set of cells would retain the original characters and remain connected with parasympathetic nerves which another set would have wholly, or mainly, newer characters and become connected with later developed nerves—the sympathetic. But this theory involves a conduction both of contractor and of inhibitory effect. The simplest case is that of the bladder of frog; contraction of this is caused both by sympathetic and by parasympathetic nerves, and so far as can be seen all the muscle fibres contract, though on sympathetic stimulation relatively feebly. The cells do not form two layers, so that there is no question of one layer being innervated by one system and the other layer by the other system. Since contraction of one part may occur without contraction of the whole, the cells, on the hypothesis, must be intermixed, and there must be conduction with a rapid decrement from cell to cell. Some evidence that there are two sets of cells in the mucous salivary glands is afforded by the unequal action on the cells of stimulating either the chorda tympani or the sympathetic, and some evidence of conduction from cell to cell is afforded by the fact that stimulation of bulbar nerves increases the secretion obtained by subsequent stimulation of the sympathetic.

> I suggested that the bulbar and sympathetic nerves may supply different gland cells in the salivary glands in 1889 (J. Physiol. 10, p. 328) on the basis of the "augmented" secretion and in 1890 (J. Physiol. 11, p. 153) on the basis of the histological changes in the glands.

There are other questions which are not yet settled. Although perhaps we need have no serious doubt that the primary action of sympathetic and parasympathetic drugs is on the cell in the region of the nerve ending, it ought not to be forgotten that there is no direct evidence that their action is not throughout the cell. The theory of localised action, so far as it is not simply a survival of the theory of an action on nerve endings, is based partly on the assumption that each cell receives sympathetic and parasympathetic nerve fibres, and partly on the results obtained in striated muscle. If each cell receives fibres from both nerve systems and one nerve ending is inhibitory and the other contractor, it seems inevitable that the affected regions of the cells on electrical stimulation must be different. But even this deduction involves an assumption of the way in which contraction and inhibition is produced; it is conceivable that contraction is produced when the nerve fibres penetrate the cell membrane, and inhibition when they remain outside it, so that the two nerves cause opposite electric charges on the cell membrane, in which case the opposed effects of electric stimulation would not show a localised action of the drugs. In striated muscle there is, however, direct evidence of localised action in the region of the nerve ending, and from the general correspondence of the action of drugs on unstriated muscle and glands to those on striated muscle, it may be inferred with a high degree of probability, that as there is localisation in one, there is also localisation in the other.

Localised action of drugs on striated muscle. The chief facts relating to this are briefly as follows:—It has long been known that the middle portion of the sartorius muscle of the frog is more excitable by induction shocks than the nerve free ends, and that a small amount of curari lowers the excitability in the middle portion without having any obvious effect on that of the end portions, i.e. that the effect of a small amount of curari is mainly, and perhaps entirely, in the region of the nerve endings. Whether curari in larger amount effects the electrical excitability of the nerve free portion has not been settled. The experiments so far made tend to show that the action of curari, though much greater in the middle region of the muscle, is not confined to it. Pollitzer (J. Physiol. 7, p. 274, 1886) stated that a large dose of curari (0·3 c.c. 1 p.c.) caused a decrease in the excitability of the nerve-free end of the sartorius of the frog, as well as in the part containing nerve endings, and Lapicque (C.R. Soc. Biol. 1, p. 991, 1906) that curari in proportion to its amount caused a decrease in the direct excitability of muscle to currents of short duration. Keith Lucas (J. Physiol. 35, p. 103, 1906) found the "optimal stimulus" of the pelvic end of the sartorius to range from 37 to 63 in the uncurarised muscle, and from 13 to 56 in the curarised muscle; the experiments, however, were not made on the same muscles, and Keith Lucas considered the excitability to be "practically" unchanged. Edström (Skand. A. Physiol. 41, p. 101, 1921) concluded from varied experiments on frog's muscle that curari in sufficient amount lowered the excitability of the general muscle substance. In some of these experiments it is possible that the diminution in excitability was due to potassium salts in the curari.

Keith Lucas, by observations on the "optimal stimuli" of the sartorius (J. Physiol. 34, p. 372; 35, p. 103, 1906) and the strength-duration curves for muscular contraction (J. Physiol. 36, p. 113, 1907), showed that currents of a certain short duration caused contraction only when applied to the nerve-containing portion of the muscle. This continned to be obtained after sufficient curari to paralyse the nerve, but ceased after a somewhat larger dose of curari. Since the excitability was local, the action of curari in suppressing it was local. He showed that a large dose of curari did not alter the form of the strength-duration curves for the pelvic end of the muscle, but, as mentioned above, his results indicated a slight decrease of excitability.

I found (Proc. Roy. Soc. B. 78, p. 186, 1906) that the tonic contraction caused by nicotine in frog's muscle was purely local on local application, and that curari prevented the effect of dilute nicotine, but not that of 1 p.c. nicotine. On microscopic examination (J. Physiol. 36, p. 347, 1907) dilute nicotine was seen only to cause tonic contraction when applied to the nerve ending region of a muscle fibre. The contracted part was spindle-shaped, and in favourable cases was seven to eight times the length of the nerve ending

(Ibid. 47, p. 163, 1913). At the end of the muscle a concentration of nicotine about 500 times as great was commonly required to cause a local contraction, and this in no long time killed the muscle. Thus, as regards the tonic contraction, the action of nicotine and of curari is more or less narrowly localised.

Boeke (Brain, 44, p. 1, 1921) describes a "periterminal" network in the sarcoplasmic sole of the nerve ending in striated muscle. He states that this network degenerates on nerve section, though somewhat later than the nerve ending itself, and suggests that it is, or contains, the receptive substance of the neural region for the quick conducted contraction.

Further, the results so far obtained do not show whether the specific excitability to drugs in the neural region is an inherited character, or whether it is produced in each generation by nerve impulses. There are facts supporting each theory. If the cells are altered in each generation by nerve impulses in such way that they become excitable to drugs we should expect the excitability to disappear in no long time when nerve impulses cease. In fact the excitability of unstriated muscle to adrenaline (and some other drugs) not only remains after nerve degeneration, but in general increases. On the other hand, Elliott found in two cats that adrenaline and hypogastric stimulation did not inhibit the bladder, and in these the sympathetic nerves to the bladder were apparently absent. The observations require confirmation, for the details given are not wholly satisfactory, but, accepting them, they suggest that the excitability to adrenaline is acquired in each generation. Here again we are met with the question whether the sympathetic nerves are connected with a separate set of cells. If they are, it is as easily conceivable that in the cases noticed by Elliott the cells did not develop as that the nerve fibres did not develop. The conception of inherited characters localised in the region of the nerve endings is not without difficulties, but some drugs have been found to have their usual action in an embryonic stage, in which it seems unlikely that nervous impulses can be sent them. Pickering (1895) found that muscarine and atropine acted on the mammalian heart in the earliest stages, and I found (1905) that adrenaline in a very early stage caused inhibition of the amnion of the chick. But Pickering did not find that muscarine had as early an action on the heart of the chick, and Kouliabko (1904) only found a trifling effect to be caused by adrenaline in one case in which he revived the excised heart of a foetal infant by perfusion; the heart, however, was apparently in a pathological state.

Classification of sympathetic and parasympathetic nerves according to pharmacological action.

H. Meyer (1912) has suggested that the cases in which there is a lack of correspondence between the effect of adrenaline and that of sympathetic nerve stimulation, and between the effect of the pilocarpine group of drugs and parasympathetic nerve stimulation, are due to the central sympathetic centre sending sympathetic fibres to parasympathetic nerves, and to the central parasympathetic centre sending parasympathetic fibres to sympathetic nerves. On this hypothesis the nerve fibres passing by way of the sympathetic chain to the uterus are partly, and those passing to the sweat glands wholly, parasympathetic fibres. This hypothesis obviously introduces great confusion into nomenclature, and there is no evidence for it beyond that which may be considered to be afforded by pharmacological action. The theory I have put forward that the similarities and dissimilarities in the action of drugs depends upon variations in the cells occurring in different periods of development is, I think, a more satisfactory explanation of the facts. But, however, this may be, it is necessary for descriptive purposes to keep to a nomenclature based on anatomy, and if it is held that there are in parts of the autonomic system two kinds of nerve fibres characterised by having a different end apparatus, this can be expressed by the understandable—though inelegant—words, cholinophil and adrenophil.

Pharmacological action has led to another extension of the term parasympathetic, in connexion with the tone of skeletal muscles, the dependence of which on the sympathetic I discuss presently. E. Frank (1920) started from the fact that atropine abolishes certain muscular tremors and tone in nervous diseases in man, and from the generally accepted statement that atropine stops the fascicular muscular contraction caused by physostigmine. He argued that since physostigmine and atropine stimulate unstriated muscle and glands innervated by parasympathetic nerves, and this effect is abolished by atropine, the similar effects of the drugs on the striated muscles must be due to their receiving parasympathetic nerve fibres. It was, however, shown by Kato and myself (1915) that whilst atropine stops physostigmine contractions which are of central origin it does not stop those which are of peripheral origin. The distinction of sympathetic and parasympathetic drugs lies in their peripheral effects, so that, on the argument the absence of effect of atropine would show the absence of nerve fibres corresponding to parasympathetic nerve fibres.

Frank also concluded on various grounds that his "pasyarmpathetic" fibres were afferent fibres conducting impulses antidromically, and that the tonic contraction they produced was in sarcoplasm. As I have said, I regard antidromic action as a function of certain afferent somatic fibres. Antidromic impulses can be set up in all nerve fibres, whether the impulses have any effect or not depends on the nature of the connexion of the fibres, and on that of the cells with which they are connected.

Schäffer (A. ges. Physiol., p. 42, 1920) supported Frank's theory on the ground that he found Tiegel's contracture in man to be increased by physostigmine and by pilocarpine, a result which might be due to an action on the central nervous system.

REFERENCES AND NOTES.

References to papers quoted in the text or in the following historical account.

Oliver and Schafer. J. Physiol. 18, p. 230, 1895.
Pickering. Ibid. p. 470.
Lewandowsky. Cntrlb. Physiol. 12, p. 599, 1898. A. (Anat. u.) Physiol., p. 360, 1899.
Boruttau. A. ges. Physiol. 78, p. 97, 1899.
Lewandowsky. Cntrlb. Physiol. 14, p. 433, 1900.
Langley. J. Physiol. 27, p. 237, 1901.
Dixon. Ibid. 30, p. 97, 1903.
Brodie and Dixon. Ibid. 30, p. 491, 1904.
Konliabko. A. di Fisiol. 2, p. 137, 1904.
Elliott. J. Physiol. 32, p. 401, 1905.
Langley. Ibid. 33, p. 374, 1905; Proc. Roy. Soc. B. 78, p. 170, 1906.
Dale. J. Physiol. 34, p. 163, 1906.
Elliott. Ibid. 35, p. 367, 1907.
Fröhlich and Loewi. A. exp. Path. 59, p. 34, 1908.
E. Frank and Isaac. Zts. exp. Path. u. Ther. 7, 1909.
H. Meyer. Deut. Zts. f. Nervenhk. 45, 1912.
Lewandowsky. Zts. ges. Neurol. 14, p. 281, 1913.

Kuno. J. Physiol. 49, p. 139, 1915.
Langley and Kato. Ibid. p. 410, 1915.
Dieden. Zts. Biol. 66, p. 387, 1916.
Dale and Richards. J. Physiol. 52, p. 110, 1918.
E. Frank. Berl. klin. Wochens, 1920. No. 31 Verh. Gesells. Deut. Nervenärtze. 10, p. 146, 1921.
Langley. J. Physiol. 55, 1921.
Luckhardt and Carlson. Amer. J. Physiol. 56, p. 72, 1921.

Degeneration of nerve endings.

Eugling (following up an observation of P. Hoffmann). A. ges. Physiol. 121, p. 275, 1908.
Langley. J. Physiol. 38, p. 504, 1909.

Historical. The early observations on Addison's disease and the results of extirpation of the suprarenal bodies led to the opinion that these bodies had a special relation to the activity of skeletal muscle. Oliver and Schafer, in their pioneer observations on the effect of extracts of the suprarenal, found that it caused contraction of small arteries, increased force of the heart beat, and had a veratrine-like action on striated muscle. These results directed attention to the action of the extracts on unstriated muscle in general. Lewandowsky (1898) found that the extract caused contraction of the unstriated muscle of the eye like that produced by stimulation of the sympathetic. He obtained (1889) no effect on the intestine or bladder. He made experiments to determine whether the action was direct on the muscle or not, a question on which there was difference of opinion. He decided for direct action, for he found that the effect of the extract persisted for some weeks after excision of the superior cervical ganglion. Boruttau (1889) noted inhibition of the intestine. Lewandowsky (1900) observed inhibition of the bladder. The action of adrenaline so far described was almost identical with that of nicotine, and the references which were made to the sympathetic were, I think, either descriptive or made as showing obviously that the effect was on unstriated muscle and not on striated muscle. Lewandowsky, to judge from a late account (1913), used "sympathetic" to include the cranial and sacral autonomic nerves. I investigated the action of suprarenal extract (1901), and found that it caused secretion from salivary and certain other glands. From its effects on a large number of tissues I concluded that it did not in any case produce the effects produced by stimulating cranial or sacral autonomic nerves, and that it did produce in nearly all cases the effects produced by sympathetic nerves. The sympathetic effects not produced by the extract were contraction of the pupil except with a larger dose, absence of preliminary contraction of the bladder of the cat, and the absence of secretion of sweat. The degree of effect on arteries I found to be, in general, proportional to the degree of sympathetic nerve effect. I confirmed and extended Lewandowsky's observation of the continuance of the effect of the extract after degeneration of postganglionic sympathetic nerves.

The action of adrenaline was to me a special case of the problem of the specific excitability of nerve endings, a theory at the time practically universally adopted. I had argued a little earlier (1901) that this theory was not true as regards the action of nicotine upon nerve cells. Thus evidence that suprarenal extract—the effects of which were so naturally explained by an action on nerve endings—did not act on nerve endings was important evidence against the general theory. Until this question was settled the exact point of action was a secondary matter. I had, however, pointed out that the stimulating and paralysing action of nicotine was not necessarily on the whole cell substance; and it was obvious from the experiments I had given on the effect of adrenaline that it must act on some special constituent of the peripheral cell. The question of the point of action was taken up by Dixon (1903). He found that apocodeine paralysed vaso-motor nerves, and that adrenaline had then no effect on blood vessels, though barium chloride still caused contraction. He concluded that adrenaline acted on the nerve endings. In the following year Brodie and Dixon, in the course of an account of the vasomotor nerves of the lungs, discussed the point of action of adrenaline and of various other drugs. They stated that the action of adrenaline was invariably that which was produced by sympathetic excitation. They upheld the conclusion arrived at by Dixon that it acted on the nerve endings, and suggested that in the experiments of Lewandowsky and myself sufficient time had not been allowed for degeneration. Brodie asked me for criticism of these results. I told him that pilocarpine caused secretion of sweat six weeks after section of the sciatic, and said there was probably something at the nerve endings which did not degenerate when the nerve endings degenerated. Brodie, thereupon, inserted a footnote in the paper, in which he abandoned the view that the action was on the visible nerve ending, and, adopting the theory of continuity, considered that the point of action was at the junction of the nerve and muscle—the myo-neural junction—and that this was in trophic connexion both with nerve and muscle. The myo-neural junction theory was only partially supported by Dixon, who continued to advocate the not improbable theory that some of the drugs mentioned in the paper acted on the anatomically visible nerve endings. Elliott (1905) gave further instances of the correspondence between adrenaline action and sympathetic action. He showed that the effect of the sympathetic on the bladder varied in different animals, and that there was a similar variation in the action of adrenaline. He found that contraction of the pupil of the dog, which a small dose of adrenaline causes, was not obtained after destruction of the mid-brain, and he showed that the action of adrenaline continued for long periods after nerve degeneration He regarded an effect or absence of effect of adrenaline as a certain test of the existence or non-existence of sympathetic innervation in unstriated muscle. His general theory, which was an elaboration of that given by Brodie and Dixon, I have mentioned in the text. In the same year it was shown by Anderson that pilocarpine caused constriction of the pupil after degeneration of the short ciliary nerves. In this year also, and in the following, in an account of the point of action of curari and nicotine on striated muscle, I put forward the theory of the existence of receptive substances, and their formation at different times in phylogenetic development.

Since that time further instances of the correspondence of the action of adrenaline and of sympathetic stimulation have been found in different vertebrates, and also instances, most of which are mentioned in the text, in which the correspondence is more or less uncertain.

The theory of the specific relation of pilocarpine and of atropine to the parasympathetic system was put forward by Fröhlich and Loewi (1904). Other bodies were gradually classed with these as parasympathetic drugs. E. Frank and Isaac (1909) and H. Meyer (1912) advocated the view that choline had a function with regard to the parasympathetic system similar to that which adrenaline is commonly held to have with regard to the sympathetic. The variations in the use of "parasympathetic" have been mentioned in the text. Dale and Richards (1918) came to the conclusion that the vaso-dilator action of adrenaline on capillaries was not correlated with any nerve system.

I have discussed the action of pilocarpine since that is the most familiar "parasympathetic" drug, but the action of choline, since it is formed in the body is the most important. The fall of blood pressure caused by choline was described by Mott and Halliburton (Phil. Trans. London, 191 B., p. 211, 1899). For the effects of acetyl-choline cp. Dale (J. Pharm. ex. Ther. 6, p. 147, 1914) and Reid Hunt (Amer. J. Physiol. 45, pp. 163, 231, 1918). Both consider the vaso-dilator action of acetyl-choline to be independent of any nerve system.

It is outside the scope of my purpose to discuss how far adrenaline and choline act normally in the body, though some reference to the action of choline on the intestine will have to be made later. But it is clear that if excessive production of either occurs it will cause a whole group of symptoms similar in the main to those caused by sympathetic or by parasympathetic nerves. Eppinger and Hess (Die Vagotonia, Berlin, 1910) state that pathological cases occur of hyperexcitability (and of hyperactivity) of the whole sympathetic system (sympathotonia), or of the whole parasympathetic system (parasympathotonia, vagotonia), and that the hyperexcitability can be detected by the increased effect of adrenaline and pilocarpine respectively. Their view that increased excitability of one is accompanied by decreased excitability of the other has not found much support.

In fish, chromaffine cells are relatively more developed than sympathetic nerve cells, and, in consequence, the theory that tissues are affected by adrenaline which are not affected by sympathetic stimulation has found some favour as concerns fish.

5. The Tissues Innervated.

All cells, except those which are specially differentiated for receiving afferent impressions, and those which migrate, may be regarded as potentially able in phylogenetic development of becoming connected with efferent nerve fibres, the final state depending upon the use to the organism. Experiment alone can determine whether such connexion has developed, or, if once developed, whether it has been retained. Cardiac muscle, nearly all unstriated muscle, most glands and some pigment cells have been shown to be affected by efferent autonomic nerves. Whether capillaries are influenced by efferent nerves is a question for discussion. But it is obvious that autonomic nerves have a different degree of effect on the different cells on which they act. The different degree of effect is very great in glandular tissue; thus nerve stimulation has a prompt and great effect on the salivary glands and little or none on the intestinal glands. And the anterior lobe of the pituitary gland which arises from the primitive pharynx has not been shown to react to nerve impulses. Unstriated muscle in most cases responds promptly and greatly to nerve stimulation either by contraction or by inhibition, but veins do not contract as readily as arteries, nor large arteries as readily as arterioles. The difference in response is partly due to the relative amount of muscle, partly to the relative degree of innervation, but it is in part due either to the muscular excitability being different, or to a difference in the ease with which nervous impulses pass to the tissue. A certain degree of lack of excitability, or of increase of nerve block, would cause the tissue to be uninfluenced by nerve impulses; it is, for example, still doubtful whether the small cerebral arteries are affected by the —apparently efferent—nerve fibres accompanying them. The question of nerve connexion with one element of a tissue and not with others arises more strikingly in the case of the black pigment cells— melanophores. The melanophores are usually considered to be connective tissue cells. There is good evidence that the melanophores of the skin in lower vertebrates are influenced by sympathetic nerve fibres, whilst the deeper lying melanophores and undoubted connective tissue cells are not known to be influenced. Apparently then nervous connexion has developed in one type only of a large class of cells. Spaeth (1916), however, chiefly as the result of observations made on the melanophores of the skin of a fish, considers that the skin melanophores are modified unstriated muscle cells. He lays stress on the fact that the clumping and dispersion of the pigment granules is brought about by reagents which cause contraction and relaxation of unstriated muscle, and on apparent transition forms between branched melanophores and the unstriated muscle of the iris. The difficulty of this view is that the histological resemblance between cutaneous melanophores and deep lying melanophores is greater than the resemblance between cutaneous melanophores and the unstriated muscle of the iris. And many observers consider that the movement of granules is due to internal movements of the protoplasm, and not as in muscle to movement in bulk. Further, if, as has been said by Lieben, adrenaline causes a slight movement of the pigment in the deep lying melanophores, the much greater response of the cutaneous melanophores might be due to the development by nerve impulses of this capacity to respond.

The details of innervation of the tissues will for the most part be considered under the separate nervous systems, but the question whether the cutaneous pigment cells, capillaries and striated muscle are innervated by any part of the autonomic nerve system will be most conveniently taken here.

Pigment cells.

Variations in the tint of the skin of lower vertebrates are brought about chiefly by the shifting of the position of melanin granules in the melanophores. When the skin is dark the melanophores are visible as branched cells, having melanin granules in their processes as well as in their cell bodies; when the skin is light the melanophores are visible as roundish black clumps. In this section we are only concerned with the evidence that nerve impulses affect directly the melanophores, so that it is unnecessary to discuss whether the concentration of the melanin granules is brought about by retraction of the cell processes or by a passage of the granules from the processes to the body of the cell without retraction of the processes, or whether movement of granules first occurs and retraction of processes later, all of which views are held. In the following account I speak of the concentration as contraction without prejudice to the question whether the cell processes are withdrawn into the body of the cell or no.

That nervous impulses either directly or indirectly cause contraction of melanophores in many lower vertebrates is shown by the facts that in appropriate conditions afferent impulses from one part of the body cause the skin of the rest of the body to become lighter in tint, that section of a cutaneous nerve causes the skin to become darker, and that stimulation of the peripheral end of the nerve causes the skin to become lighter. The promptitude and the constancy with which an effect is produced by the different forms of experiment vary in different animals and in different species of the same animal. They are also influenced by local conditions which act directly on the melanophores, and which tend to cause them to contract or expand, such as variations in moisture, pressure, light, the presence of chemical bodies, and probably temperature. In general, nerve section appears to have a prompter and greater effect on fish than on the frog, whether this is due to the normal tone being greater in fish is uncertain.

The contraction of melanophores caused by nervous action has been taken by nearly all observers to be due to a direct action on them, but some have attributed it to an indirect effect of change in blood supply. It was shown by Lister (1858) on severance of all the tissues of the thigh of a frog except the sciatic nerve—the artery being cut between two ligatures—that the skin became pale. Biedermann (1892) noted that the expansion of the melanophores on ligaturing the artery might begin in a few minutes, and become maximal in $\frac{1}{4}-\frac{1}{2}$ an hour, and that on ligature of the vein the effect was produced more slowly. Both observers, however, considered that the nerves had a direct action. This view is based rather on a balance of probability than on conclusive experiment. The conclusive experiment would be to show that nerve stimulation causes contraction of the melanophores when there is no change in blood supply, and when no metabolic products which can cause contraction are formed. The experiment which most nearly fulfils these conditions in the frog is that made by Königs (1915). Königs states that by stimulating the sciatic with currents of a certain voltage, duration and rhythm, there is contraction of melanophores without alteration in blood flow, but as the frogs were not curarised the possibility of an action of muscular metabolic products on the melanophores is not definitely excluded, though it is true that the experiment of Dutartre mentioned below and other similar experiments makes this unlikely. In several accounts it is implied that nerve stimulation in the frog causes darkening of the skin more quickly than anaemia, but accurate data are wanting. In fish the effects of anaemia have been less studied, but so far as the observations go they suggest a separation of pigmentomotor and vaso-motor effect. Pouchet (1875) found that section of the inferior maxillary nerve in the turbot caused contraction of melanophores and that ligature of the inferior maxillary artery did not. Barbour and Spaeth (1917) describe the melanophores of Fundulus heteroclitus removed from the body as long remaining expanded in Ringer's fluid. On the other hand v. Frisch (1911) states that anaemia in fish causes great pallor, and it is known that the melanophores contract after death.

The pigmento-motor nerve fibres in all probability belong entirely to the sympathetic system. Pouchet (1876) found in the turbot that section of the ventral division of a spinal nerve distal of its ramus communicans caused darkening of the skin in the area supplied by the division, and that section centrally of the ramus did not. The result was confirmed by Rynberk (1906) in the sole, and by v. Frisch (1911) in the minnow. Dutartre (1890) found in the frog that stimulation of the upper part of the spinal cord after section of all the rami communicantes on one side did not cause on this side change of tint of the skin. Possible antidromic action has not been looked for.

The theory of direct action is to some extent supported by the microscopical observations of Ballowitz (1893) and others, who describe nerve endings on the melanophores of fish; but nerve endings are also described on the cells of the choroid plexus, on which nerve action is unknown, and on capillaries on which it is doubtful. Lieben (1906) found that adrenaline applied to the surface of the frog's web caused contraction of the melanophores without change of blood flow, and Barbour and Spaeth (1917) that it caused contraction of the melanophores of Fundulus after removal from the body.

From the results mentioned above, taken as a whole, it may be fairly concluded that nerve fibres cause contraction of melanophores by direct action.

Carnot (1896) put forward the hypothesis that the melanophores are supplied with inhibitory as well as with contractor fibres. The basis for this was apparently that some substances, e.g. amyl nitrite caused expansion of the melanophores of the frog. Whilst this fact is no evidence for the hypothesis, there are other facts which are at least suggestive. Pouchet described the melanophores in the turbot after nerve section as not being completely expanded, and

Barbour and Spaeth observed that adrenaline applied after ergotoxine to the melanophores of excised scales of Fundulus caused expansion instead of contraction. It may be noted that they found slight expansion to be caused by pilocarpine.

REFERENCES AND NOTES.

Lister. Phil. Trans. London, 148, p. 607, 1858.
Pouchet. J. de l'Anat. Physiol., p. 113, 1876.
Dutartre. C. R. Acad. Sci. 111, p. 610, 1890.
Biedermann. A. ges. Physiol. 151, p. 455, 1892.
Ballowitz. Zts. wiss. Zool. 56, p. 673, 1893 (nerve endings in Teleostean melanophores).
Carnot. C. R. Soc. Biol., p. 927, 1896.
Lieben. Zntrlb. Physiol. 20, p. 108, 1906.
Rynberk. Ergeb. Physiol. 5, p. 429, 1906.
v. Frisch. A. ges. Physiol. 133, p. 319, 1911.
Königs. Étude de l'excit. d. nerfs, etc., Paris, 1915.
Spaeth. J. Exp. Zool. 20, p. 293, 1916.
Barbour and Spaeth. J. Pharm. exp. Ther. 9, p. 356, 1917.

A full summary of the numerous papers on pigment cells up to 1906 is given by Rynberk, see also Königs. For papers on fish melanophores, 1906 to 1911, see v. Frisch, and for later papers Ballowitz (A. ges. Physiol. 157, p. 165, 1914). Many of the observations deal with the adaptation of tint to surroundings. Adaptation has a less direct bearing on the question of the existence of pigmento-motor fibres than nerve section and stimulation, since it involves controversy as to the degree of direct action of light on the melanophores; in consequence I have not thought it necessary to discuss it.

The red chromatophores which occur in some animals are influenced by nerve stimulation in the same way as the black chromatophores. It is doubtful whether there is any nerve effect on yellow chromatophores or on guanin cells; concentration of particles in both cells when exposed to light has been described by Ballowitz.

Sympathetic nerves and the innervation of capillaries.

There is no doubt that constriction and widening of capillaries can be caused by chemical substances independent of variations in the blood pressure, and there is little doubt that this may be produced by an action on the epithelioid cells forming their walls. Since nerve fibres, in part of sympathetic origin, run from the small arteries and accompany the capillaries, it is a natural hypothesis that the sympathetic controls the size of the capillaries in the same way that it controls the size of the small arteries. The evidence of such action is, however, very slight. The only instance in which capillary contraction, certainly independent of pressure variation, has been obtained on nerve stimulation is that found in the nictitating membrane of the frog by Steinach and Kahn (1903). But the effect they obtained was a longitudinal folding of the epithelioid wall, and they attributed it to contraction of cells outside the capillaries, probably they considered the cells described by Rouget (1873) and by S. Mayer (1886), and not to a contraction of the epithelioid wall. Rouget and Mayer described branched cells, which they held to be muscle cells surrounding certain capillaries. It is possible that some capillaries, especially in a certain stage of development, have scattered muscle cells in their outer coat, but there is no evidence that this is of more than a local occurrence in amphibia, and none that it occurs in fully formed mammalian capillaries. In any case, Steinach and Kahn's observations afford no support to the theory that the sympathetic send efferent fibres to the epithelioid wall of capillaries.

Many observers have watched the capillary circulation in the web of the frog during sciatic stimulation, and no one has found that the stimulation causes any certain contraction in them. The difference in the effect on the arteries of the web and that on the capillaries and veins when the sympathetic nerve fibres supplying it are stimulated is most striking. The arteries contract and stop the blood flow completely or nearly completely, whilst the capillaries remain fairly full; sometimes, however, the capillaries become empty of blood corpuscles and their diameter decreases. This was noted by Lister (1858) on stimulating the spinal cord. He attributed the effect to the arterial contraction being sufficient to stop the blood corpuscles, but insufficient to stop the plasma. I found (1911) on prolonged (and repeated) stimulation of the sympathetic, or of the sciatic nerve after curari had been given, that sometimes the capillaries gradually became empty instead of remaining partly full. These variations in capillary diameter are, I think, more probably caused by variations in the balance of pressure inside and outside the capillaries and by the action of metabolites than by direct nerve action. The capillaries are surrounded by lymph spaces, and increased pressure in these would tend to compress the capillaries. The conclusion arrived at by Roy and Graham Brown (1879), that capillary diameter is little influenced by variations in arterial pressure, is no doubt true for the circulation in the web of the frog in ordinary conditions. The long-known fact that in the web great contraction of the arteries is not accompanied by any considerable diminution in the diameter of the capillaries shows that the normal pressure distends them but little, and that the extra-vascular pressure is slight. But I do not agree with the statement of Krogh (1920) that arterial pressure has no appreciable effect on capillary diameter; it has seemed to me that there is a distinct, though slight, effect both when pressure falls owing to local contraction of arterioles, and when it rises, owing to contraction elsewhere. And I take it that there is at times a local increase in extra-capillary pressure capable of considerably compressing the capillaries. However, this may be, it is certain that if any contraction of capillaries is produced by nerve stimulation in the web of the frog it is of quite a different order from that which the nerves produce on the arterioles.

In the mammal the facts are different. It has long been noticed that stimulation of the cervical sympathetic usually causes the ear, with the exception of the veins, to become completely pale, so that the capillaries as well as the arteries are emptied of blood. This has generally been taken as showing that when the arterial contraction cuts off the blood supply, the pressure in the tissue spaces has been sufficient to empty the capillaries. Rouget (1879), however, considered that the pallor of the face caused in man by emotions could only be due to capillary contraction. In recent years there has been a revival of inquiry into the conditions of capillary circulation. Dale and Richards have shown that histamine causes intense dilatation of the capillaries. Cannon pointed out that in shock there was stagnation of blood in the capillaries. Krogh found local dilatation on mechanical stimulation of the tongue of the frog, and the old observation of the red stripe caused by stroking the skin has received renewed investigation. These evidences of independent action of the capillaries, though all in the direction of vaso-dilation, have revived attention to the question of vaso-constrictor nerves for them, and Hooker (1920) interprets the pallor of ear (cat) caused by sympathetic stimulation as involving active capillary contraction. No satisfactory evidence is, however, adduced to show that this constriction is not passive. It will be noticed that if the pallor in the tissues of the mammal is caused in part by contraction of the capillaries, the action of nerves on them must be nearly as prompt as that on the arterioles.

The question of the production of capillary dilatation by afferent nerve fibres will be dealt with in the section on antidromic action. The action of adrenaline has been referred to above (p. 32).

REFERENCES AND NOTES.

Lister. Phil. Trans. Roy. Soc., Pt. 2, p. 607, 1858.
Rouget. A. Physiol. norm. et path 5, p. 656, 1873.
Roy and Graham Brown. J. Physiol. 2, p. 338, 1879.
Rouget. C. R. Acad. Sci. 88, p. 916, 1879.
S. Mayer. Wien. Sitzb. 93. Abt. 3, p. 45, 1886; Anat. Anz. 21 p. 442, 1902.
Steinach and Kahn. A. ges. Physiol. 97, p. 105, 1903.
Langley. J. Physiol. 41, p. 483, 1911.
Hooker. Amer. J. Physiol. 44, p. 30, 1920.
Krogh. J. Physiol. 53, p. 405, 1920.

Capillary contractility. The theory that the capillaries receive vaso-constrictor nerves from the sympathetic takes for granted that the capillaries are contractile. The direct evidence of capillary contractility, except in response to chemical substances, is limited to a few tissues chiefly in amphibia. The discussion of the question is of long standing. Allen Thomson in 1835 (Todd's Cyclopædia Anat. and Physiol. 1, p. 671) discussed the evidence and supported the theory. In the subsequent 30 years most observers inclined to the opinion that the variations known to occur in the diameter of the capillaries were due either to variations in blood pressure or to variations in the elasticity of the wall caused by inflammatory processes and by reagents. A change of opinion began with Stricker's observations in 1865 (Wien. Sitzb. 51, Abt. 2, p. 16). He found spontaneous variations in diameter in the capillaries of the nictitating membrane of the frog and (ibid. 52, p. 379, 1866) contraction in them on stimulating with strong induction shocks. Similar stimulation also caused contraction in the capillaries of the tail of young tadpoles, but there was usually no effect in fully grown tadpoles, and very rarely in the capillaries of the mesentery of the frog. Golubev (A. mik. Anat. 5, p. 49, 1869) and Tarchanov (A. ges. Physiol. 9 p. 407, 1874) confirmed the constriction of the lumen of the capillaries of the nictitating membrane of the frog on electrical stimulation, but attributed it to a swelling of spindle-shaped elements, later shown to be nuclei. Stricker (Wien. Sitzb. 74, Abt. 3, p. 313, 1876) confirmed his previous results, and added that strong induction shocks caused contraction of capillaries in fully grown tadpoles after they had been treated with dilute alcohol. He observed sometimes the bulging of the spindle-shaped elements, as described by Golubev, but noted contraction of the whole tube in young tadpoles. Severini's papers (1878-9) I have not been able to obtain; he is quoted by numerous writers as having shown that bulging of the nuclei, and often narrowing of the lumen of the whole capillary is caused by oxygen and the reverse change by carbonic acid. Roy and Graham Brown (1879) considered that the diameter of the capillaries was regulated by metabolites probably acting directly on the capillary wall. The opinion that the capillaries were contractile had been supported by Rouget, but the contractile part he held were branched muscle cells surrounding the capillary tube. Such cells he described as present in the hyaloid membrane of the frog (A. Physiol. norm. et path., p. 601, 1873). Later (1879) he found similar cells around the capillaries of the tail of the tadpole, the capsulo-pupillary membrane of new-born mammals, and of the omentum of young mammals. All these, he said, contracted on electrical stimulation. S. Mayer (1886, 1902) confirmed Rouget's observations, which had received little attention, and described branched cells as being present around the capillaries of the intestine and of the bladder of the salamander and frog. Some at any rate of these cells were almost certainly connective tissue cells. Steinach and Kahn (1903), on direct stimulation of the capillaries of the pericœsophageal membrane of the frog and of the omentum of the kitten, did not obtain contraction in capillaries of less than about 10μ in diameter.

Krogh (J. Physiol. 52, p. 457, 1919) has given reason to believe that in resting striated muscle, some of the capillaries are too small to admit blood corpuscles, but that they dilate during contraction and on massage. In general, he holds that the capillaries are constantly undergoing active changes in diameter.

In mammals ocular evidence of contractility is given by mechanical stimulation (see Ebbecke. A. ges. Physiol. 169, p. 1, 1917).

Sympathetic nerves and the innervation of striated muscle.

In recent years there has been much discussion on the question of the innervation of striated muscle by sympathetic nerve fibres. Bremer, in 1882, found that occasionally in limb muscles and frequently in the muscles of the tongue, nerve endings were formed by non-medullated as well as by medullated nerve fibres. Some of the non-medullated fibres were fibres which had lost their medulla in their course in the muscle, some were branches of medullated fibres, but some arose from a non-medullated plexus, and could not be traced further centrally. The nerve endings of the two kinds of fibres differed in appearance; they were commonly, but not always, close together, in the muscle fibre. Perroncito[1] (1902) put forward the theory that the muscle fibres were supplied with nerve endings both from the sympathetic and from the cerebro-spinal systems. Mosso (1904) chiefly on the basis of Perroncito's observations, suggested that the sympathetic governed the tone and slow contraction of muscle, and the cerebro-spinal

[1] Quoted from Mosso. Perroncito's paper, in Arch. Ital. de Biol. 38, p. 393, 1902, does not give this theory.

(i.e. somatic) nerves governed quick contraction; both forms of contraction he attributed to the muscle fibrils. The question of the connexion of the sympathetic with muscle fibres was brought to the front by Boeke in a series of observations from 1909 onwards. Boeke (1912, 1917), and subsequently Agdur, and Boeke with Dusser de Barenne (1919), showed by the degeneration method that a plexus of sympathetic nerves was in fact present in muscle, and that from this plexus nerve fibres proceeded to the muscle fibres forming endings such as had previously been described in normal muscle.

In these observations the somatic nerves of the eye, finger, and intercostal muscles were cut peripherally of their trophic centres, and after the nerves had degenerated the muscles were treated by the silver impregnation method. The results obtained by Boeke (1912, 1917) on the eye muscles are not easy to interpret. He found that the endings of the great majority of the non-medullated fibres were not degenerated a few days after intercranial section of the motor nerves, but were degenerated in about three weeks. He considered that these fibres were autonomic fibres issuing in the nerve roots. If this is so, they are different from other autonomic fibres, either in having no peripheral nerve cells on their course, or in degenerating peripherally of the nerve cells. In view of this difficulty rigid proof is required that the non-medullated nerves in question do not arise from the medullated fibres, and become impregnated with silver in consequence of the persistence of the sheath (cp. p. 26). Against this possibility it may be urged the Boeke describes the endings as being hypolemmal, and that hypolemmal fibres are not known to have a sheath. The evidence of the existence of sympathetic nerve endings in the eye muscles was not very decisive; it depended on the presence of a few non-medullated fibres after degeneration of the somatic nerves, and on an apparent slight decrease of non-medullated fibres after degeneration of the sympathetic nerves.

It may be mentioned that Botezat (Zts. wiss. Zool. 84, 1906; Anat. Anz. 35, 1910) described the muscles of birds as having nerve endings both from medullated and non-medullated nerves; he did not, however, trace the central origin of the latter.

Since the sympathetic supplies the vessels of the muscles, it is obvious that a plexus of sympathetic fibres will remain after degeneration of the somatic nerves. The essential point is, whether fibres from this plexus form nerve endings under the sarcolemma of the muscle fibres, or whether they all end in connexion with blood vessels. Boeke, Agdur, and Dusser de Barenne describe hypolemmal nerve endings, though Agdur takes some to be epilemmal. But nerve endings, definitely sympathetic, have so far been only shown by the silver impregnation method, and if all muscles have them it is remarkable that they have not been described in methylene blue preparations of the sartorius of the frog in which a certain number of nerve endings of somatic nerves and of nerve fibres around blood vessels are brought out with ease and certainty. Whilst this apparent contradiction has still to be cleared up, the evidence at present is decidedly in favour of Boeke's views.

The histological results led de Boer to investigate the function of the sympathetic nerves supplying the muscles. He found (1913) that section of the rami communicantes of the nerves of the leg of the frog caused loss of tone of the muscles, and he concluded that the function of the sympathetic was to keep up muscular tone. Pekelharing and Hoogenhuyze had brought forward much evidence to show that muscular tone involved an increase of creatine, i.e. protein breakdown. The tone they considered, was brought about by somatic nerves, and on the lines of Bottazzi's theory that it might be sarcoplasmic. De Boer, holding the tone to be caused by sympathetic nerves, suggested that the sympathetic caused protein breakdown in sarcoplasm. de Boer, a little later in the same year, found that extirpation of the lumbar ympathetic in the cat caused a loss of tone in the muscles of the hind limb and of the tail. The results obtained by subsequent observers have varied greatly, but on the whole they show that a slight decrease of muscular tone may be caused, both in the frog and in the mammal, by section of the sympathetic. Observations in addition to those mentioned below are given in the notes at the end of this section.

The theory of de Boer that the tone of the muscles is entirely due to impulses passing by the sympathetic is not tenable. Previous experiments on the effects of cutting the sympathetic and of excision of the ganglia had, I think, given ample evidence that neither operation caused complete loss of tone in the muscles of the head and limbs. Dusser de Barenne (1913), Negrin y Lopez and v. Brücke (1917) and v. Rijnberk (1917) showed that decerebrate rigidity is not dependent on sympathetic innervation. Jansma (1915) described in the frog greater loss of tone on cutting the sciatic than on cutting the rami communicantes, and Dusser de Barenne (1916) a greater loss of tone on cutting the posterior roots than on cutting the rami. De Boer (1915) referred the decerebrate rigidity occurring after section of the sympathetic to a tetanic tone caused by the somatic nerves, but tetanic tone, if it involves the fibrillae, causes increased production of CO_2, and, according to Roaf, the CO_2 production in decerebrate rigidity is not perceptibly greater than that of the curarised muscles. Further, Lilljestrand and Magnus (1919) found that the tone caused by tetanus toxin, shown by H. Meyer to be due to central stimulation, continues after section of the sympathetic.

When a loss of tone has been obtained by sympathetic nerve section, its duration has not been found to be the same by different observers. Dusser de Barenne (1916) found that the loss of tone on excision of the lumbar sympathetic in the cat disappeared in four to eight weeks. Negrin y Lopez and v. Brücke (1917), in similar experiments, found the loss of tone (when it occurred) disappeared in a day or two. Ducceschi (1919), who observed a slight drooping of the ear of the rabbit after excision of the superior cervical ganglion, noted that the position of the ear became normal in a few weeks, but that by weighting the two ears a slight loss of tone could be detected for six months.

Whilst the theory that the sympathetic causes some degree of muscular tone is satisfactory in that it provides a function for the nerve endings described by Boeke, the experiments which have been made do not decide definitely whether the loss of tone consequent on sympathetic section is, or is not, due to concurrent vaso-dilation. The variation in degree and permanence of effect on tone in the mammal comes well within the limits of the variations which have been found in vascular effects. It is true that de Boer (1915) has described a decrease of tone in the gastrocnemius of the frog on cutting the rami communicantes after excision of the abdominal viscera, when presumably the circulation was at most slight, but in view of the varying results which have been obtained in the frog, further more decisive observations are required.

A defect in the evidence that the sympathetic causes tonic contraction in muscle is that no one has been able to obtain contraction by stimulating it. Even the rectus abdominis and the arm muscles of the frog, which are so prone to tonic contraction, show no visible change on stimulation of the rami communicantes. If, then, the sympathetic causes tone in striated muscle, the nervous impulses producing it are different from the nerve impulses which cause tone in unstriated muscle. The lack of effect of nerve stimulation is the more striking in that in mammals tonic contraction is readily produced by stimulating the cerebellum.

> The question of the nature of the changes occurring in tonic contraction is outside the scope of the subject discussed in this section, but there is a possibility regarding it which bears on the question of sympathetic innervation. It is known that there are some forms of continuous contraction set up by the central nervous system in which the electric state of the tissue remains constant, and in which apparently there is no chemical change. It is conceivable that this form of contraction should be produced by a continuous current from the nerve cells which gives rise to a static charge on the membrane of the contractile substance. On this hypothesis, on the one hand the absence or occurrence of tonic contraction on nerve stimulation would depend on the character of the nerve, and on the other hand the probability that the tonic contraction is due to the fibrillae and not to sarcoplasm would be very great.

In view of the theory of Pekelharing that muscular tone is accompanied not by a breakdown of carbohydrates, but by a breakdown of protein with formalin of creatine, and of de Boer's theory that tone is maintained by the sympathetic, observations have been made to determine whether cutting off impulses coming by the sympathetic decreases the amount of creatine in muscles. Jansma (1915) found a slight decrease in the creatine content of the muscles of the leg of the frog, on cutting the rami communicantes of the sciatic plexus, but the results hardly seem to be outside the limits of experimental error. Riesser (1916) paralysed the motor nerves in rabbits with a minimal dose of curari, cut the sciatic on one side, and then subjected the animals to conditions tending to stimulate the sympathetic heat centre postulated by H. Meyer. He found less creatine on the side on which sympathetic impulses had been cut off by severance of the sciatic, but this only occurred if the circulation through the limb was restricted by tying the femoral arteries. The conditions in these experiments are too complex to carry conviction. On the other hand, v. Rijnberk (1917) induced decerebrate rigidity in cats after excision of the lumbar sympathetic on one side, and found no difference in the creatine content of the muscles of the hind legs on the two sides, and Kahn (1919) found less creatine in the arm muscles of the frog during the clasping season than in the leg muscles.

Experiments have also been made on the effect of the sympathetic on the oxygen use and carbonic acid formation in muscle. The starting point of these was the observation of Frank and Voit that curari given to dogs in not much more than the amount required to paralyse motor nerves did not cause increased CO_2 production, provided the muscles were at rest, but did, as in Pflüger's experiments, when given in larger quantities. The larger quantity Frank and Voit considered paralysed the sympathetic vaso-motor nerves. Mansfeld and Lukács (1915) estimated the respiratory exchange in dogs, and found that in a lightly curarised animal, section of the sciatic caused a decrease of respiratory exchange if the sympathetic was intact, but did not do so if the sympathetic connexions had been cut. On their view the section abolished muscular tone, and so decreased the respiratory exchange of the muscles.

Section of the sympathetic would, however, tend to decrease the amount of blood in the viscera, and so tend to decrease the respiratory exchange. An increase in CO_2 formation would tend to connect the sympathetic with carbohydrate metabolism, but no decrease of muscle glycogen was found by Ernst (1915) on stimulation of the sciatic in lightly curarised frogs. Further, Nakamura (1921) determined the oxygen use in the muscles of the lower part of the leg and obtained no change in it on section of either sympathetic or sciatic in anaesthetised and in decerebrated and lightly curarised cats. The existence of tone before section of the nerves was not, however, directly tested. Stimulation of the sympathetic caused a decrease in oxygen use; since this was accompanied by a great decrease in blood flow, the presumption is that it was caused by the decrease in oxygen supply, although an inhibitory action was not definitely excluded. There is evidence that several forms of muscle tone, and possibly all, involve chemical change at the onset only. So far as this occurs little or no change in metabolism would be expected to occur on abolishing it.

E. Frank (1920), in connexion with his theory of the production of tone (plastic tone) by antidromic impulses (cp. p. 52), considers that the sympathetic instead of increasing tone inhibits it. The evidence given is that adrenaline abolishes certain forms of muscular twitching and tone in man, that it abolishes in the dog the awkwardness and rigidity caused by injecting physostigmine, and that Schäffer found Tiegel's contracture to be abolished by injecting adrenaline. Schäffer took ergographic tracings of the contraction of the muscles of the forearm in a man in which Tiegel's contracture was marked. He inferred that the abolition of the contracture by adrenaline was not due to vaso-constriction since the effect was produced in two to four minutes, and in a plethysmyograph experiment made at another time the decreased volume of the arm only began in $5\frac{1}{2}$ minutes; and, further, since adrenaline had a somewhat quicker effect than cutting off the blood supply by compression of the upper arm. The theory assumes that the action of adrenaline is in all cases on the end apparatus of sympathetic nerves, and in the particular instances that the effect produced by adrenaline is not produced either by central action or by peripheral vaso-constriction. In Schäffer's experiments a greater decrease of blood supply would be caused by adrenaline than by compression of the upper arm. As mentioned above (p. 33), several observers have been unable to find that adrenaline has any effect on normal muscle.

It has been suggested that the pseudo-motor action of the chorda tympani on the tongue described by Vulpian and by Heidenhain indicates that parasympathetic fibres may form nerve endings in connection with striated muscle. The pseudo-motor effects in this and other regions is, I think, more probably due to an action of metabolites on muscle altered by section of its motor nerve. The facts bearing on the question will be discussed later.

We may summarise the results as follows:—There is a balance of histological evidence that sympathetic nerve fibres form hypolemmal nerve endings on some striated muscle fibres, but the evidence is insufficient to show that they form nerve endings on all of them. There is no satisfactory evidence that the sympathetic directly affects the formation of creatine, or CO_2 in muscle. The only change in muscle which has been found on stimulating the sympathetic is a decrease in oxygen use, an effect which may be due to the concurrent decrease of oxygen supply. There is a balance of evidence that section of the sympathetic nerves supplying a muscle causes a slight loss of tone, but conclusive evidence that this is not the result of section of vaso-motor nerves is lacking. Until this question is settled it is premature to discuss whether the sympathetic acts on sarcoplasm or on fibrillae; it appears, however, to be certain that no form of tonic contraction on striated muscle attributed to sympathetic nerves is distinguishable from that produced by somatic nerves.

REFERENCES AND NOTES.

Bremer. A. mik. Anat. 21, p. 165, 1882.
Perroncito. Ges. med. Ital. 1902 (quoted from Mosso).
Mosso. A. Ital. de Biol. 41, p. 183, 1904.
Boeke. Intern. Monats. Anat. Physiol. 28, p. 377, 1911.
   Verhd. Anat. Gesells, 1912.
Roaf. Quar. J. exp. Physiol. 5, p. 31, 1912.
Boeke. Anat. Anz., p. 343, 1913.
de Boer. Folio Neurob. 7, pp. 358, 837, 1913.
Dusser de Barenne. Ibid. p. 651.
de Boer. Zts. f. Biol. 65, p. 239, 1915.
Jansma. Ibid., p. 365.
Mansfeld and Lukács. A. ges. Physiol. 161, p. 467, 1915.
Ernst. Ibid., p. 483.
Dusser de Barenne. Ibid. 166, p. 145, 1916.
Riesser. A. exp. Path. u. Pharm. 80, p. 183, 1916.
Negrin y Lopez and v. Brücke. A. ges. Physiol. 166, p. 55, 1917.
Rijnberk. A. néerl. de Physiol. 1, p. 727, 1917.
Boeke. Verband. k. Akad. v. Wetens, Amsterdam. Sect. 2, 1917
Agdur. Proc. R. Acad. Amsterdam 21, p. 930, 1919.
Boeke and Dusser de Barenne. Ibid., p. 1227.

Ducceschi. A. di Fisiol. 17, p. 59, 1919.
Kahn. A. ges. Physiol. 177, p. 294, 1919.
Liljestrand and Magnus. Münch. med. Wochens, p. 551, 1919.
E. Frank. Berl. klin. Wochens., p. 725, 1920. Deut. Zts. Nervenheilk. 70, p. 146, 1921.
Schäffer. A. ges. Physiol. 185, p. 42, 1920.
Nakamura. J. Physiol. 55, p. 100, 1921.

Further references are given by Boeke, op. cit. 1911, de Boer, op. cit. 1915, and Dusser de Barenne, op. cit. 1916.

Kuno (J. Physiol. 49, p. 139, 1915) did not find any loss of tone in the muscles of the leg of the frog on cutting the rami communicantes running to the sciatic nerve. Salek and Weilbrecht (Zts. f. Biol. 71, p. 247, 1920), in similar experiments, but in which the hind legs were suspended in cold water, observed as a rule no loss of tone at first, but later there was sometimes a slight loss. Cobb (Amer. J. Physiol. 4, p. 478, 1918) found no loss of tone in the hind legs and tail of the cat on cutting the sympathetic chain between the 4th and 5th lumbar ganglia. v. Rijnberk (A. néerl. 1, p. 727, 1917) extirpated the lumbar sympathetic in cats, and some time later induced decerebrate rigidity; he found no constant difference in tone on the two sides.

Kure, Hiramatsu, and Naito (Zntrlb. Physiol. 28, p. 130, 1914) stated that the tone of the diaphragm depended upon the splanchnic nerves. Mansfeld (A. ges. Physiol. 161, p. 478, 1915) described a loss of tone on cutting the sciatic of the frog after curari had been given in quantity just sufficient to paralyse the motor nerves.

The second rise in the curve of a single muscular contraction and the prolongation of the contraction of a veratrinised muscle have both been held by some observers to be sarcoplasmic. de Boer (op. cit. 1915) considered both to be due to impulses passing by the sympathetic. On stimulating the sciatic of a curarised and veratrinised muscle he obtained in some cases a slow prolonged contraction without preliminary quick contraction. de Boer (op. cit. 1915) also claimed that section of the lower rami communicantes in the frog delayed rigor mortis; as is known, cessation of circulation in the spinal cord leads to a general stimulation of the sympathetic. de Boer held that sympathetic nerves running to muscle were similarly stimulated, and by causing increased metabolism hastened rigor. Dusser de Barenne (op. cit. 1916), however, contested the statement that section of the sympathetic hastens rigor mortis.

Note on the action of adrenaline and of pilocarpine on the sweat glands.

As mentioned on page 29, Dieden found that local injection of adrenaline into the cat's foot caused secretion in certain conditions. The conditions were nerve section or very deep anæsthesia. The absence of secretion in other conditions he considered was due to inhibitory nerve impulses passing (probably) by the posterior roots. In the text, I have taken the action of adrenaline solution as being due to the fluid in which the adrenaline is dissolved. The reasons for this were (1) that when 0.1 p.c. adrenaline caused secretion, Ringer's fluid, locally injected, caused a greater secretion; (2) that the secretion caused by adrenaline (when obtained) was strictly local, and did not occur in the foot beyond the injected area; and (3) that 1 p.c. barium chloride, amyl nitrite, hypertonic salt solution, or distilled water, also commonly caused a secretion. Unlike Dieden, I did not find that nerve section had any constant effect on the result of local injection. All secretion, including that caused by barium chloride, was prevented by atropine.

After the earlier sheets of this book had been printed I made further experiments, and in some of these, adrenaline (generally 0.001 p.c.) caused markedly more secretion than Ringer's fluid. Whether this was due to a different depth of injection, or to a difference in the excitability of the glands, or to an occasional action of adrenaline remains to be determined.

The action of pilocarpine after the nerves have degenerated does not depend on the fluid in which it is dissolved (cp. p. 41) for the secretion is not confined to the part of the foot injected, but spreads over the whole of the secretory surface. The time after denervation at which pilocarpine still causes secretion varies greatly in different animals. This, indeed, was shown by the early observations of Marmé and of Luchsinger (Hermann's Hdb. d. Physiol. v., Th. 1, p. 428, 1883), who occasionally, but only occasionally, obtained secretion 2–3 weeks after section of the sciatic. I may note that normally there is great variation in the excitability of the sweat glands in different cats, occasionally pilocarpine and nerve stimulation cause no secretion. Local injection of any fluid, but especially of adrenaline, reduces the response of the glands to pilocarpine.

Printed by
W. Heffer & Sons Ltd.,
Cambridge, England.

Made in United States
Troutdale, OR
04/03/2024

18905851R00056